Quantum Information Science - Recent Advances and Computational Science Applications

Edited by René Steijl

Published in London, United Kingdom

Quantum Information Science - Recent Advances and Computational Science Applications
http://dx.doi.org/10.5772/intechopen.1001631
Edited by René Steijl

Contributors
Abdulbast Abushgra, Carlos Pedro Gonçalves, Christian Jansson, Claudio Sanavio, Fernando Javier Gómez-Ruiz, Ferney Rodríguez, Luis Quiroga, Neil F. Johnson, René Steijl, Rutuja Kshirsagar, Sauro Succi

First published in London, United Kingdom, 2024 by IntechOpen
IntechOpen is the global imprint of INTECHOPEN LIMITED, registered in England and Wales, registration number: 11086078, 167-169 Great Portland Street, London, W1W 5PF, United Kingdom

British Library Cataloguing-in-Publication Data
A catalogue record for this book is available from the British Library

Additional hard and PDF copies can be obtained from orders@intechopen.com

Quantum Information Science - Recent Advances and Computational Science Applications
Edited by René Steijl
p. cm.
Print ISBN 978-0-85466-575-4
Online ISBN 978-0-85466-574-7
eBook (PDF) ISBN 978-0-85466-576-1

For EU product safety concerns:
IN TECH d.o.o., Prolaz Marije Krucifikse Kozulić 3, 51000 Rijeka, Croatia,
info@intechopen.com or visit our website at intechopen.com.

Meet the editor

Dr. René Steijl works as a senior lecturer in aerospace engineering at the James Watt School of Engineering at the University of Glasgow. His main areas of research are the development of novel, multiphysics numerical methods for flow simulations and the development of quantum algorithms for computational fluid dynamics. In the past two decades, his work has included large-scale investigations of helicopter aerodynamics, and high-speed aerodynamics, as well as the development of new mathematical models for rarefied gas flows. In recent years, he has pioneered quantum algorithms based on the concept of lattice-based modeling of fluids.

Contents

Preface IX

Section 1
Computational Science and Engineering Applications 1

Chapter 1 3
Quantum Computing for Simulation of Fluid Dynamics
by Claudio Sanavio and Sauro Succi

Chapter 2 21
Floating-Point Arithmetic with Consistent Rounding on a Quantum Computer
by René Steijl

Chapter 3 45
Robust Quantum Algorithms for Early Fault-Tolerant Quantum Computing
by Rutuja Kshirsagar

Section 2
Quantum Communication and Security 59

Chapter 4 61
Quantum Key Distribution Approaches
by Abdulbast Abushgra

Chapter 5 79
Vulnerability of Quantum Information Systems to Collective Manipulation
by Fernando Javier Gómez-Ruiz, Ferney Rodríguez, Luis Quiroga and Neil F. Johnson

Section 3
Data Representation and Quantum Information Processing 103

Chapter 6 105
Unitary Maps and Quantum Artificial Neural Networks
by Carlos Pedro Gonçalves

Chapter 7 **131**
An Alternative Approach to Probability in Quantum Information Science
by Christian Jansson

Preface

In recent years, quantum computing research has become a very active and diverse area of research. There has been a notable increase in investigations into applications to computational science and engineering with the potential for significant speed-up over simulations on classical (nonquantum) computers. This development formed the main motivation for the current book.

The seven chapters in this book present advances in the development of quantum algorithms for applications in computational science and engineering (first section) and applications of quantum information processing to data security (second section). In the final section of this book, a chapter focusing on unitary maps and quantum artificial neural networks is followed by a chapter presenting in detail an alternative approach to probability in quantum information science.

The application of quantum computing to numerical simulations of fluid dynamics has become an important research topic in recent years. In Chapter 1, the authors first provide a detailed perspective of these recent developments. This perspective highlights important remaining research questions and challenges. As explained by the authors, a key challenge is the inherent nonlinearity of the governing equations of fluid dynamics. Various approaches are actively being researched to address this challenge. One such approach is based on the concept of Carlemann linearization. In the second part of Chapter 1, the authors detail how this linearization approach can be used in the context of quantum algorithms based on the lattice Boltzmann method.

The work presented in Chapter 2 was also motivated by ongoing work related to developing quantum algorithms for fluid dynamics simulation. In computational fluid dynamics applications, but also in the wider context of computational science and engineering applications, it can be observed that quantum algorithms often need to perform arithmetic operations implemented in terms of quantum circuits. In the majority of quantum algorithms where such operations occur, the arithmetic operations for real data take place in terms of a fixed-point representation based on integer arithmetic. For scientific computing, a floating-point representation of real data is more appropriate, based on evidence of classical computing. Chapter 2 summarizes key features of a reduced-precision floating-point data representation developed specifically to minimize the number of qubits needed in a quantum circuit implementation. Then Chapter 2 details two different applications of this data representation relevant to a wide class of computational engineering. A particular focus of this chapter is the implementation of the quantum floating-point arithmetic such that the result is consistently rounded to the nearest real value that can be presented by the chosen data format. As explained by the author in this chapter, this type of rounding is crucial for applications where governing equations are used that represent conservation laws for mass, momentum, and energy.

The first two chapters in Section 1 described work in quantum algorithm development targeting quantum computers with a higher level of fault tolerance than that present in the current and near-future hardware, usually referred to as the near-term intermediate-scale quantum (NISQ) era. With an increased level of fault tolerance, quantum computing hardware moves toward fault-tolerant quantum computing (FTQC). A key question that motivated Chapter 3 is how to transition from NISQ to FTQC with regard to quantum algorithm development. One idea discussed in this chapter is the definition of an additional era in quantum computing preceding the era of FTQC, an era sometimes referred to as early fault-tolerant quantum computing (EFTQC). For algorithms belonging to EFTQC, a key challenge is to establish a trade-off between hardware and algorithm parameters such that quantum circuits that are still limited in depth and width are used effectively on hardware with a limited amount of error correction. For two example algorithms (phase estimation and randomized Fourier estimation), this is worked out in detail.

It can be argued that quantum communication, especially the quantum key distribution (QKD), is the most mature and most ubiquitous application of quantum technology. QKD holds great potential in providing security in data communication by using key features of quantum mechanics. Quantum key distribution is the topic of Chapter 4. As detailed by the authors, the practical implementation of QKD faces several challenges that hinder its widespread adoption. Following a brief introduction to quantum communication and its advantages over classical communication, the authors then describe key components of QKD, with key challenges to its wide adoption presented first. Chapter 4 ends with a presentation of prospective solutions to the outlined challenges.

Chapter 5 addresses a number of key aspects to consider when concepts such as a global quantum Internet and quantum Internet of Things come closer to reality. As explained in detail by the authors, an important question for society concerns such systems' vulnerabilities. A new form of vulnerability in such systems that the authors identify based on detailed many-body quantum mechanical calculations, namely vulnerability to collective manipulation, forms the main topic of this chapter. The authors purposely structured the discussion in this chapter so that the first sections are self-contained and can be read by nonspecialists, while all the necessary details of the many-body quantum mechanics and model calculations are presented in later sections.

Unitary quantum maps represent a bridge between classical and quantum dynamical systems theories. When applied to quantum artificial neural networks, unitary quantum maps allow one to address these networks as quantum networked dynamical systems. In Chapter 6, the authors describe the application of unitary maps to quantum artificial neural networks. Their work includes simulations of these networks on a cloud-based quantum computer (IBM Q Experience). Chapter 6 ends with a summary of the implications of the presented work for quantum complexity research, quantum chaos theory, and quantum computing in more general terms.

In the final chapter (Chapter 7), the author describes a probabilistic framework for formulating classical probability theory, quantum probability, thermodynamics, diffusion, and the Wiener integral using a set of four axioms or principles. As pointed out by the author, the work presented is not to be understood as an alternative interpretation of quantum mechanics. In Chapter 7, the framework introduced is applied to several well-known experiments. At the end of this chapter, a brief summary of the work in terms of quantum information processing is provided.

In summary, it is clear that the seven chapters cover a wide range of contributions in a very active area of research. We hope that the readers will enjoy reading the chapters and will find them useful and stimulating.

René Steijl
James Watt School of Engineering,
University of Glasgow,
Glasgow, United Kingdom

Section 1

Computational Science and Engineering Applications

Chapter 1

Quantum Computing for Simulation of Fluid Dynamics

Claudio Sanavio and Sauro Succi

Abstract

The implementation of quantum algorithms for the simulation of classical fluid dynamics poses a fundamental challenge due to the nonlinearity of the fluid equations. In this work, we provide a pedagogical introduction to quantum computing algorithms for simulating classical fluids, with a special focus on the Carleman-Lattice Boltzmann algorithm, which has captured significant attention in the last couple of years. While this algorithm demonstrates satisfactory convergence to analytical solutions for systems at low-to-moderate Reynolds numbers, it also shows an exponential depth of the corresponding quantum circuit. As a result much further analysis is needed to assess the availability of the Carleman-Lattice Boltzmann method on a quantum computer.

Keywords: quantum computing, fluid-dynamics, carleman-linearization, lattice-boltzmann-method, collision operator, streaming operator

1. Introduction

Quantum computing has been vigorously pursued in the recent years, mostly in view of the spectacular speed-up it may provide as compared to the performance of electronic computers, for a number of scientific applications [1]. Quantum computing can be traced back to Richard Feynman's epoch-making paper, in which he observed that physics "ain't classical," hence it ought to be simulated on quantum computers [2]. Following Feynman's observations (with due credit to previous investigators such as T. Toffoli and E. Fredkin), early theoretical work on quantum computing was performed in the 1980s, for example, Deutsch's work on the link between quantum theory, universal quantum computers and the Church–Turing principle [3]. Then, with the publication of Shor's algorithm for integer factoring and Grover's search algorithm in the middle of the 1990s, the research area gathered significant momentum in terms of theoretical work and quantum computing hardware as well. The research area of quantum computing has continued to grow ever since [4–6].

In terms of applications, the simulation of quantum many-body systems has received special attention, due to its scientific and industrial relevance, as well as due to its close link with quantum hardware, meaning the possibility of mapping quantum Hamiltonian directly into native quantum gates. In this chapter, we shall focus on a much less beaten track, namely the use of quantum computers for the simulation of classical fluids.

To this purpose, it is convenient to refer to the physics-computing plane defined by the following four quadrants: CC: *Classical computing for classical physics*; CQ:

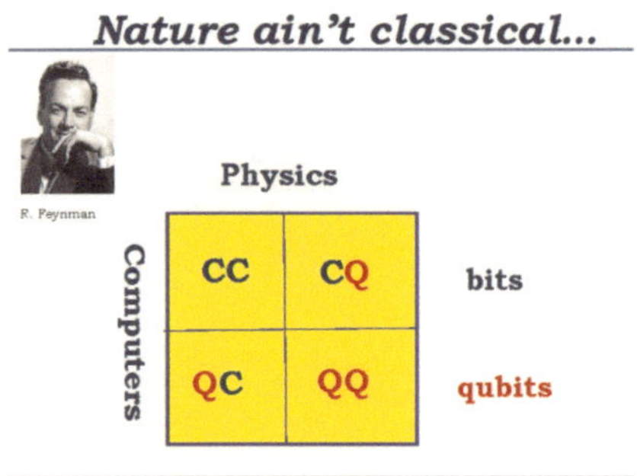

Figure 1.
The four quadrants in the physics-computing plane. The CC and CQ quadrants are the mainstay of current scientific computing. QQ is the quadrant invoked by Feynman, and QC is the quadrant relevant to this paper.

Classical computing for quantum physics; QC: *Quantum computing for classical physics*; QQ: *Quantum computing for quantum physics* as shown in **Figure 1** (taken from [7]).

Feynman's observation pertains to the CQ sector shown in **Figure 1**, where one is often confronted with exponential complexity barriers associated with the phase-space of quantum many-body problems [8, 9]. The basic idea is that such exponential barriers can be handled by the corresponding exponential capacity of qubit representations offered by the QQ quadrant. In this chapter, we shall focus on the opposite off-diagonal QC quadrant, where the power of quantum computing might be brought to the fruition of hard computational problems in classical physics. In particular, we shall focus on computational fluid dynamics (CFD), which presents a ceaseless quest for better and more efficient algorithms and computers, mostly on account of the problem of fluid turbulence [7, 10]. The physics of fluids presents two general features which are not shared by quantum physics: *nonlinearity* and *dissipation*. Both set a major challenge to quantum computing, which draws much of its power from the superposition principle and the unitary and Hamiltonian structure of quantum mechanics. In the following, we shall present a few potential strategies to deal with both issues above.

The chapter proceeds as follows. In Section 2, we introduce the main advantages and issues in dealing with the QC sector. To address some of these challenges, various hybrid classical-quantum approaches have been proposed. These strategies leverage the strengths of both classical and quantum computing, offloading certain tasks to the classical computer while utilizing the quantum computer for operations where it offers a significant speed advantage. In Section 3, we provide a concise review of key hybrid approaches from the literature. While not exhaustive, this review highlights the most prominent approaches. In Section 4, we review some of the fully quantum approaches, that is, algorithms where the computation is completely performed by the quantum computer, and only the extraction of the information is due to the classical computer. Sections 2, 3 and 4 are mostly based on the works of the authors published in Ref. [7]. In Section 5, we present and comment on one of the most promising among these approaches: the Quantum Carleman-Lattice Boltzmann algorithm, originally developed by the authors in Ref. [11]. In Section 6, we draw our conclusions and perspectives.

2. Challenges facing quantum computational fluid dynamics (QCFD)

As mentioned in the introduction, realizing the potential of quantum computing means leveraging distinctive features of quantum mechanics that, by definition, are not available on classical computers. However, it also follows that it is precisely these specific features that expose major challenges in realizing simulations with a quantum advantage.

The main quantum mechanical concepts spawning the potential benefit of quantum algorithms are *quantum superposition* and *quantum parallelism*. The quantum state in a Q-qubit coherent register can be described by the Schrödinger wave function, defined by 2^Q complex amplitudes for 2^Q states in superposition. The square of each of these amplitudes defines the probability of finding the system in the corresponding state after *quantum measurement*. By encoding classical data in terms of these amplitudes an exponential saving in storage can be achieved when the number of qubits is compared to the required number of classical bits.

Let us illustrate the idea for the specific case of turbulent flow simulation. Turbulence features a Re^3 complexity, where the Reynolds number Re represents the relative strength of convection (nonlinearity) versus dissipation. Most real life problems feature Reynolds numbers in the many-millions; for instance, an airplane features $Re \sim 10^8$, implying $O(10^{24})$ floating-point operations per simulation. This is basically the best one can afford on a nearly-ideal Exascale classical computer. The simulation of regional atmospheric circulation flows takes us at least another two decades above in the Reynolds number, hence totally out of reach for any foreseeable classical computer [12, 13]. On the other hand, the minimum number of qubits Q required to represent Re^3 complexity can be roughly estimated as $2^Q = Re^3$, namely:

$$Q = 3Log_2 Re \sim 10\, log_{10} Re \tag{1}$$

This simple estimate shows that 80 qubits match the requirement of full-scale airplane design, while $O(100)$ qubits would enable regional atmospheric models [12] or nuclear fusion applications [14]. However, several key challenges stand in the way of this task.

First, quantum measurement: Extracting classical information collapses the quantum wave vector into a single eigenstate (a house of cards). Hence, to get classical values for each of the amplitudes, multiple realizations of the quantum state vector are needed with an associated set of measurements.

Second, conditional operations: In the quantum circuit model, the "classical" information is not available to the quantum gate operations performed in the circuit. Specifically, gate operations can be conditional on the state of one or more control qubits, while specifying gate operations conditional on one or more of the complex amplitudes defining the wave function is impossible. Therefore, when classical data are encoded in terms of amplitudes, a rotation angle in a quantum gate operation, this information is required at the time the circuit is compiled. This angle cannot be changed at run time as a function of "classical" data encoded in quantum amplitudes.

Third, time marching: In an algorithm involving multiple iterations or time steps, the overhead associated with quantum measurements used to extract classical data and re-initialization of the quantum state for the next iteration, scale quadratically with the grid size [15] and consequently they severely tax the quantum CFD efficiency. It should be observed that the well-known Harrow-Hassidim-Lloyd (HHL) [16] algorithm for linear system solution assumes that the input and output data are encoded in terms of quantum amplitudes without including the cost of setting up the quantum state and extracting the classical solution.

Fourth, circuit depth: With the exception of quantum measurement operations, quantum mechanical operators are unitary, linear and reversible; hence, they can be directly implemented on a series of unitary quantum gates (quantum circuit model). Usually, in order to quantify the computational cost, producers and specialists take track of the number of two-qubit gates employed in the circuit. In fact, two-qubit gates, such as the CNOT gate, come in the hardware with an error that is several orders of magnitude (depending on the platform) larger than single-qubit gates errors. We know from the DiVincenzo theorem [17] that the number of two-qubits gates needed to implement a generic Q qubits unitary operation scales as 4^Q, thus voiding the exponential advantage of the embedding process. In order to sidestep this issue, the unitary operation that we want to perform in our circuit has to be *combinatorially block diagonal* [18]. Hence, this provides a stringent constraint on the possible algorithms that can actually be implemented to simulate classical fluids.

If we further assume that the velocity field is represented in terms of *amplitude encoding* [19], that is, the components of velocity vector at all grid points are represented in terms of the amplitudes defining the wave function, then the *no-cloning theorem* prevents the use of (temporary) copies of any of these amplitudes. So, evaluation of u^2 or $u\frac{\partial u}{\partial x}$ cannot be performed by storing a temporary copy $temp = u$, to perform the computation of the value of u^2 as $u \times temp$. Also, for data encoded in terms of the complex amplitudes of the Schrödinger wave function, there is a need for this state vector to have a unit norm, since these amplitudes represent probabilities of states. This means that for an operator attempting to compute the squares of these complex amplitudes, the resulting state vector loses unitarity. So, even without the no-cloning theorem complicating such a step, this points to a further problem with computing nonlinear terms. This loss of unit norm of the quantum state vector represents an example of a *non-unitary operation*, typically associated with a corresponding loss of information. A similar argument runs for orthogonality of the eigenstates, since a nonlinear propagator rotates the state by an angle which depends on the state itself.

Fifth, dissipation: Dissipation breaks hermicity of the quantum propagator. This can be recovered by augmenting the system with its Hermitian conjugate, so that the doubled system is Hermitian by construction. One can dispense with doubling degrees of freedom by representing the non-unitary update as a weighted sum of unitaries. This introduces a failure probability which needs to be handled probabilistically, that is, by executing the update conditional on the successful measurement of an ancilla qubit.

To summarize, dealing with nonlinear and dissipative terms raises major extra challenges for quantum computational fluid dynamics (QCFD) besides the ones commonly encountered for the simulation of quantum systems.

3. Hybrid quantum/classical approaches

The challenges sketched above have resulted in the fact that most of the existing work in QCFD is based on a *hybrid quantum/classical approach*, with the quantum processor performing computations for which efficient quantum algorithms exist, while the output is then passed on to classical hardware to perform further computational tasks not (yet) amenable to quantum algorithms.

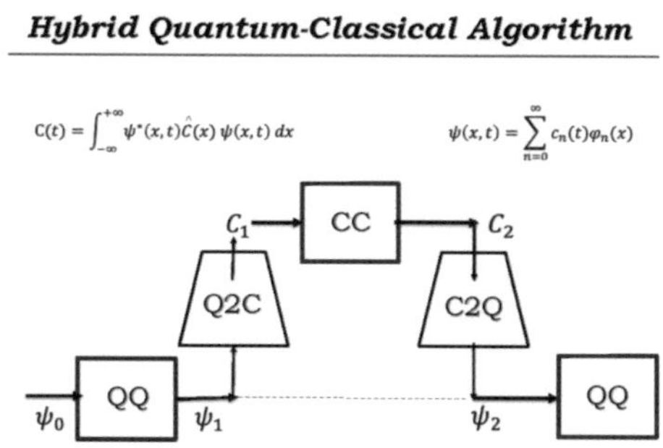

Figure 2.
Sketch of a hybrid quantum-classical algorithm. The illustration shows the steps involved in a single time step or single iteration, including preparation of the subsequent step/iteration.

Figure 2 illustrates the idea. The quantum state ψ_0 is advanced to the next quantum state ψ_1 *via* a QQ algorithm. The quantum state ψ_1 is then measured to generate classical observables C_1 which are advanced to C_2 by a CC algorithm. The classical observables C_2 are then used to reconstruct the quantum state ψ_2, ready for the next QQ step. The Q2C conversion shown in **Figure 2** involves quantum measurements and requires repeated measurements over a statistical sample of quantum states, since none of them can be re-used. The C2Q reconstruction requires the preparation of all the quantum eigenstates, hence a full reset of the quantum circuits from scratch. Both operations impose a substantial computational burden on the hybrid algorithm, typically scaling exponentially with the number of qubits.

Examples of previous works using the *hybrid quantum/classical approach* include the works of Steijl [20], Gaitan [21] and Budinski [22]. The algorithm presented by Gaitan uses Kacewiz's quantum amplitude estimation ODE algorithm [23] as applied to the set of nonlinear ODE's resulting from standard discretization of the Navier-Stokes equations. As shown, for certain "non-smooth" problems (illustrated using the quasi-1D flow in converging-diverging duct with normal shock wave), the complexity analysis shows potential for exponential speed-up, so that the challenges associated with hybrid quantum/classical computing can potentially be overcome. For the linear advection-diffusion equation, Budinski [24] presented a quantum algorithm based on the lattice Boltzmann method [25, 26]. The algorithm can perform multiple successive time steps with no need for quantum measurements and re-initialization of qubit register between the time steps if suitable re-scaling of solution vector is applied to deal with non-unitarity. In the quantum circuit model, the fact that velocity field is unchanged between successive time steps implies that the operator implementing $u\,\frac{\partial u}{\partial x}$ can be re-used in multiple time steps. Budinski [22] then extended the work to Navier-Stokes equations in stream function-vorticity formulation. Then, velocity field updates during each time step means that quantum circuit implementation of convection terms cannot be re-used during multiple time steps and that the classical value of u (as well as other flow field data) is needed to define quantum circuit implementation for the next time step. This shows that it is the nonlinearity that forces the use of a hybrid quantum/classical approach, similar to previous work where Navier-Stokes equations were discretized, for example, the hybrid quantum/classical algorithms for fluid simulations based on quantum-Poisson solvers [20].

In summary, key challenges for the *hybrid quantum/classical approach* are: (i) cost and complexity of (repeated) measurements; (ii) statistical noise due to sampling of the quantum solution; (iii) cost and complexity of (repeated) re-initialization.

The development of more efficient (re-)initialization techniques appears to be a key factor for the future success of quantum computing for fluids.

4. Quantum fluid dynamics strategies

For the sake of concreteness, let us refer to the Navier-Stokes equations for time-dependent incompressible flows:

$$\frac{\partial \mathbf{u}}{\partial t} + \mathbf{u} \cdot \frac{\partial \mathbf{u}}{\partial \mathbf{x}} = -\frac{\partial P}{\partial \mathbf{x}} + \nu \Delta \mathbf{u} \tag{2}$$

$$\nabla \cdot \mathbf{u} = 0 \tag{3}$$

where $\mathbf{u}$ is the velocity vector, P is the pressure, defined for location $\mathbf{x}$ as function of time t. The kinematic viscosity is defined by ν and density has been conventionally set to a unit constant value. Eq. (3) enforces mass conservation, while Eq. (2) is based on momentum conservation in each coordinate direction. Eq. (2) and Eq. (3) highlight that it is the convection term that represents the nonlinearity, that is, the second term on the left-hand side of Eq. (2). Writing the Navier-Stokes equations in non-dimensional form, such that $\mathbf{x}$ and $\mathbf{u}$ are scaled by reference length L_{ref} and U_{ref}, respectively, it follows that in Eq. (2) the term ν is replaced by $1/Re$, where Reynolds number $Re = U_{ref}L_{ref}/\nu$. For Stokes flow, that is, with Reynolds approaching 0, nonlinear terms are vanishingly small, but still not to be neglected since they are responsible for non-trivial long-range correlations especially important in biological flows [27]. For high Reynolds number (turbulent) flows, obviously the nonlinear terms play the leading role.

Different strategies have been adopted to simulate fluid dynamics on quantum computer, mostly involving hybrid algorithms. Since they have been reviewed in the recent perspective [7], PNAS [28], and their time complexity has been analyzed thoroughly in [29], here we simply list them, directing the reader to the original literature.

4.1 Nonlinear quantum ODE solvers

This approach consists in treating the discretized Navier-Stokes equations as a set of nonlinear ODEs and advancing them in time by developing bespoken quantum nonlinear ODE solvers. This approach has been pioneered by Gaitan [21] and demonstrated for the case of a Laval nozzle on grids with $O(30-60)$ grid points over 1400 time steps. The critical issue, though, is that the concrete implementation of the quantum oracle is left to a classical computer.

4.2 Nonlinear variational quantum eigenvalue solvers

It has recently been proposed that variational quantum eigenvalue (VQE) solvers, a major tool of the QQ sector, might be extended to the fluid equations as well [30]. The basic idea is to use quantum computers for the efficient generation of variational eigenfunctions by suitable circuit parametrization. The associated energy functional is then minimized through a classical procedure. The appeal of these methods is the possibility to import algorithmic know-how from quantum mechanical applications, especially quantum chemistry. To date, however, there is no evidence that the procedure can meet the standards of accuracy required by computational fluid dynamics.

4.3 Functional approach

It is long known that any nonlinear problem can be mapped into a linear one in a space of higher dimensions, a technique also known as Carleman embedding or Carleman linearization [31]. The Carleman embedding trades nonlinearity for infinite dimensions and nonlocality. An alternative procedure to recover linearity is to take a probabilistic approach and formulate a functional kinetic (Liouville) equation for the probability distribution function (PDF) of the fluid velocity field. Formally [32]:

$$\partial_t P[\mathbf{u}] + \frac{\delta}{\delta \mathbf{u}}(f(\mathbf{u})P[\mathbf{u}]) = 0 \tag{4}$$

where $P = P[\mathbf{u}]$ is the functional PDF and $\dot{\mathbf{u}} = f(\mathbf{u})$ is the equation of motion (hence $f(\mathbf{u})$ is a nonlinear and non-local operator in function space). This is linear by construction, regardless of the nonlinearity of the dynamics. Once the fluid equations are discretized on a grid with G lattice nodes, the Liouville distribution P_G becomes a $3G$-variate PDF $P_G(\mathbf{u}_1 \ldots \mathbf{u}_G)$ which lives in a $O(3G)$-dimensional space. For large-scale flow applications, G can reach values up to multi-billions, hence fully under the "curse of dimensionality."

Fortunately, for most practical purposes, low-order marginals prove often capable to retain the essential physical information. Marginalization is known to introduce coupling terms between different marginals, thus raising the need for appropriate closures expressing the high-order marginals as functionals of the low-order ones. This is a classical topic in non-equilibrium statistical physics, which may draw significant benefits from modern developments in tensor networks theory [33, 34].

The idea can be taken one level further by noting that Eq. (4) is linear, but still not unitary. In Ref. [35] the authors presented a method to relate the classical evolution and the unitary evolution of a quantum system, making use of the Carleman-Koopman embedding, which transforms the Liouville equation into an equivalent many-body Schrödinger equation.

This is formally very appealing, but still subject to the dimensional curse; even in quantum chemistry nobody knows how to handle molecules with billions atoms with ab initio methods. In addition, Giannakis et al. [35] show that for chaotic systems, the spectrum of the Hamiltonian contains continuous modes that are pretty hard to tame.

Always in the spirit of functional methods, in Ref. [29], the authors develop another form of Koopman-von Neumann approach based on the level-set method as applied to classical nonlinear field theories [29]. The formalism is applied to hyperbolic PDE's and Hamilton-Jacobi equations, but its extension to the Navier-Stokes equations is addressed only marginally, making it difficult to draw firm conclusions.

4.4 Carleman embedding

Introducing indices m and n to denote interacting neighbors, and using Einstein summation, the fluid equations can be recast in terms of a linear operator $\mathcal{L}$ and quadratic operator $\mathcal{Q}$, as follows,

$$\frac{du_l}{dt} = \mathcal{L}_{lm}u_m + Re\, \mathcal{Q}_{lmn}u_m u_n \tag{5}$$

The pressure term was removed for simplicity, although this all but a minor item, since pressure-flow-stress coupling is going to affect the structure of the Carleman matrix.

Discretizing velocity derivatives introduces a key challenge: Carleman linearization creates a growing hierarchy of variables ($U_{lm} = u_l u_m$). Each level's tensor has $O(G\kappa^{k-1})$ components, κ being the sparsity of the $\mathcal{Q}$ matrix, and spreads across a larger spatial neighborhood. This nonlinear uplift incurs dimensionality and locality costs.

Despite this, Liu et al. [36] presents a promising quantum Carleman linearization algorithm for the 1D Burgers equation showing a polylog scaling with the number of grid points, that is an exponential improvement over classical approach. However, for a given time span T, the algorithm shows a complexity $T^2 Poly(log T)$. The authors simulated shock formation at surprisingly high Reynolds numbers (up to 40), exceeding theoretical limits and warranting further analysis.

The key question then is: Is there a maximum Reynolds number above which the Carleman procedure fails to converge? Also, when it does converge, how many Carleman levels are required to reach the desired accuracy at a given Reynolds number?

The answer may well depend on the chosen representation of the fluid equations and in the following we turn to a promising candidate in this respect, namely the Lattice Boltzmann method [25, 37].

5. Carleman-lattice Boltzmann

In a recent paper [38], Cheung and co-workers argued that Lattice Boltzmann might be more convenient for Carleman linearization, since the fluid nonlinearity is formally encoded in the Mach number instead of the Reynolds number, which makes of course a huge difference, as the Mach number is typically many orders of magnitude smaller than the Reynolds number (for an airplane Ma ~ 1 vs. Re $\sim 10^8$). They performed a classical Taylor-Green Vortex (TGV) simulation based on a Carleman-Lattice Boltzmann scheme, showing excellent agreement at moderate Mach number, with just three Carleman levels [38]. It is an unsteady flow of a decaying vortex which has an exact closed form solution of the incompressible Navier-Stokes equations. The initial conditions for the three velocity components are:

$$u_x = A_x \cos(k_x x) \sin(k_y y) \sin(k_z z), \tag{6}$$

$$u_y = A_y \sin(k_x x) \cos(k_y y) \sin(k_z z), \tag{7}$$

$$u_z = A_z \sin(k_x x) \sin(k_y y) \cos(k_z z). \tag{8}$$

The TGV is a crucial benchmark in CFD because it allows a direct comparison with a known analytical solution. More recently, in Ref. [11], the authors showed the convergence of the Carleman method to the analytical solution of a Kolmogorov-like flow, a close 2D analogue of the TGV with periodic initial conditions:

$$u_x = A_x \cos(k_x y) \tag{9}$$

$$u_y = A_y \cos(k_y x). \tag{10}$$

It was shown that the Carleman system truncated at second order for Moderate-to-low Reynolds number, up to 100 (see **Figure 3**) provided excellent agreement up to few hundred time steps.

Let us remind that the Lattice Boltzmann (LB) method is based on the following scheme: In equations:

$$f_i(\mathbf{x} + \mathbf{v}_i \Delta t, t + \Delta t) - f_i(\mathbf{x}, t) = -\frac{f_i - f_i^{eq}}{\tau} \tag{11}$$

where $f_i(\mathbf{x}, t)$, is the probability to find a representative fluid parcel with discrete velocity $\mathbf{v}_i$ at position $\mathbf{x}$ and time t. In the above, f_i^{eq} is the corresponding discrete local equilibrium and τ is a local relaxation time, fixing the viscosity of the lattice fluid. Importantly, the local equilibrium is a quadratic function of the Mach number $Ma = u/c_s$, c_s being the speed of sound. A defining feature of the LB method is that it allows to use a fixed, generally small, number of discrete velocities $\mathbf{v}_i$. For instance, it turns out that just nine discrete velocities are sufficient to fully recover the physical properties of a two-dimensional flow defined on a square lattice. The nine velocities are depicted in **Figure 4** and described in **Table 1**. The model is called D2Q9. A key feature of LB, not shared by the macroscopic representation of fluid dynamics, is that streaming proceeds along straight lines, defined by the discrete velocities. This

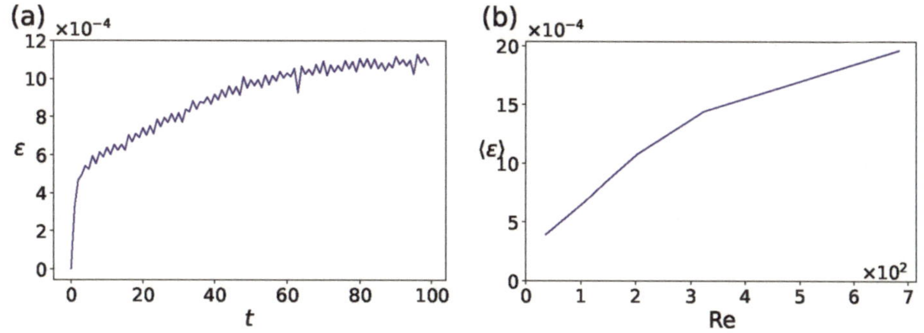

Figure 3.
(a) The deviation ε in time of the second order Carleman solution from the original LB solution for the case of a Kolmogorov-like flow at Re ~ 10, on a 32 × 32 grid. (b) The mean deviation ⟨ε⟩ as a function of the Reynolds number for the same flow.

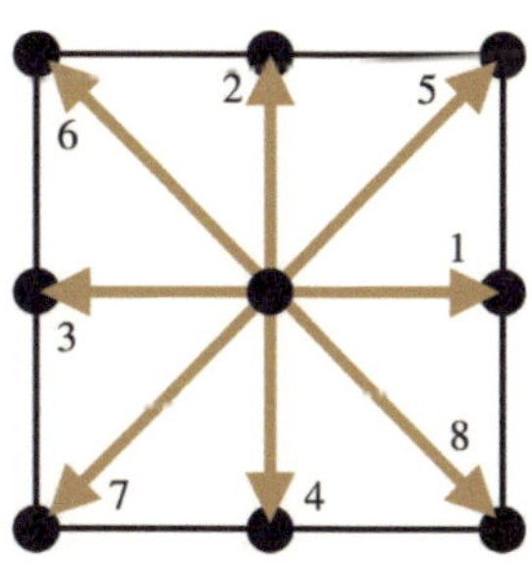

Figure 4.
The D2Q9 model, defined on a two-dimensional square lattice, with the corresponding set of nine velocity vectors labelled by index i, cf. ***Table 1****.*

i	0	1	2	3	4	5	6	7	8
c_i	$(0,0)$	$(1,0)$	$(0,1)$	$(-1,0)$	$(0,-1)$	$(1,1)$	$(-1,1)$	$(-1,-1)$	$(1,-1)$

Table 1.
The velocity set of the D2Q9 model.

operation is *exact*, in that it does not involve any floating-point calculation but just a memory shift from the lattice position $\mathbf{x}$ to another lattice position $\mathbf{x} + \mathbf{v}_i \Delta t$. Also to be observed that collisions are completely local, hence perfectly amenable to parallel computing. Because of this, dissipation is an emergent property which does not require any second-order spatial derivative, as it is the case for the Navier-Stokes equations. Besides their mathematical elegance, these properties are at the roots of the computational efficiency of the method and its amenability to parallel computing.

A key point of using the discrete-velocity Boltzmann formalism instead of Navier-Stokes is that, owing to the double dimensional phase-space, in the latter nonlocality (streaming) is linear while nonlinearity (collision) is local, while in Navier-Stokes the two merge into a single $\mathbf{u}\nabla\mathbf{u}$ convective term.

On classical computers, this disentanglement proves extremely beneficial because of the local nature of the collision term, which makes the system highly amenable to parallel computation. On quantum computers instead it is not simple to relate together the streaming and the collision steps, as we are going to show next. When writing the Lattice Boltzmann functions $f_i(x,t)$ in a quantum state we can employ amplitude encoding, using a quantum register $|\cdot\rangle_v$ for embedding the discrete velocities v_i and a second quantum register for encoding the lattice sites $|\cdot\rangle_x$. The fluid is then described by the following quantum state

$$|\psi(t)\rangle = \sum_i \sum_x f_i(x,t)|i\rangle|x\rangle. \tag{12}$$

Since the collision step is local and depends only on the velocities of the fluid in the lattice site, one may apply an operator $\hat{\Omega}$ which performs the collision step on the quantum register adapted to the velocities and another operator $\hat{S}$ encoding the streaming process to be applied to all quantum registers, being it both non-local and velocity dependent. This ideal circuit is shown in **Figure 5(a)**. It turns out that since $\hat{\Omega}$

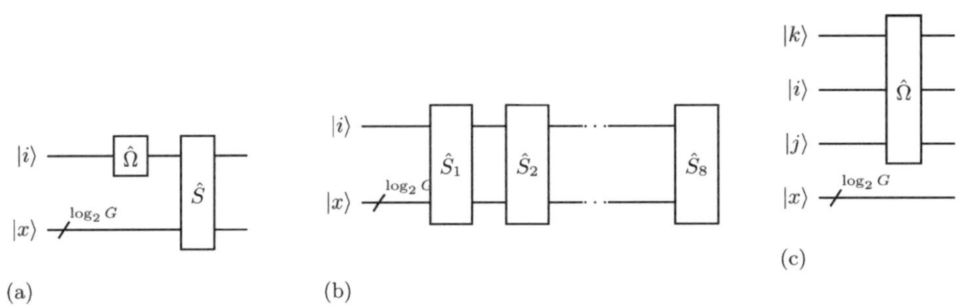

Figure 5.
The sketch of three quantum circuits for different lattice Boltzmann implementations for a grid with G lattice sites for the D2Q9 model. In (a) the collision is applied on the velocity register, whereas the streaming is a global operation applied on both the position and the velocity registers. In (b) the evolution is obtained as a series of streaming in the eight velocities of the D2Q9 model. In (c) the quantum register $|k\rangle$ holds the information about the truncation order, whereas $|i\rangle$ and $|j\rangle$ are the velocities associated to the Carleman variables.

is a nonlinear and non-unitary operator, this circuit cannot be realized, unless specific symmetry conditions hold, such as those used in Ref. [39]. Unfortunately, these do not conform to the universal quantum computation paradigm.

We may then opt for hybrid strategies whereby only streaming (collisions) would be performed on a quantum computer, leaving collisions (streaming) to a classical one. Interestingly, this leads to different circuits, as sketched in **Figure 5**. **Figure 5b** and **c** shows the circuits for collision-free streaming and streaming-free collision processes, that we are going to discuss next.

5.1 Collision-free streaming

The quantum computing algorithm for streaming is based on the approach used by Steijl and co-workers [40]. The streaming is a non-local process shifting the particles in the next neighboring site, depending on the corresponding velocity, without affecting their value. Thus, this step can be executed by moving the probability function $f_i(\mathbf{x}, t)$ to $f_i(\mathbf{x} + \mathbf{v}_i, t + 1)$ at the next time step.

The corresponding operation on the quantum state reads as follows:

$$|\psi(t)\rangle = \sum_i \sum_{\mathbf{x}} f_i(\mathbf{x}, t)|i\rangle|\mathbf{x}\rangle \xrightarrow{\hat{S}} |\psi(t+1)\rangle = \sum_i \sum_{\mathbf{x}} f_i(\mathbf{x}, t)|i\rangle|\mathbf{x} + \mathbf{v}_i\rangle \tag{13}$$

That is, only the positions register is affected by the streaming process, whereas the velocity register acts as a control. The streaming operator is obtained as a combination of the streaming in the different directions, for the D2Q9 model with eight velocities (plus the null velocity) is $\hat{S} = \oplus_{i=1}^{8} \hat{S}_i$, that are therefore applied depending on the value of the qubit in the velocity register. The streaming can be obtained by applying a series of multiqubit controlled operations on the position register. We show in **Figure 6** an example for the $\hat{S}_1$ streaming operator when applied on a 8×8 lattice.

5.2 Streaming-free collision

A completely different approach is employed to perform the collision step: Carleman linearization. Based on the Lattice Boltzmann method, Itani et al. [41] developed an approach termed *Carleman for second-quantized Lattice Boltzmann*. This terminology stems from the fact that the Boltzmann operator, defined by:

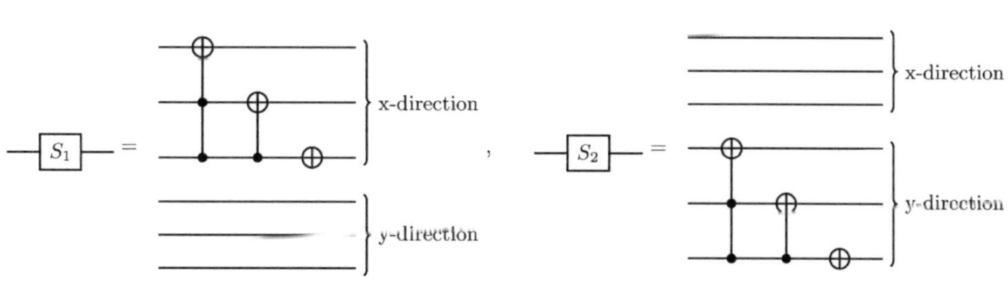

Figure 6.
*The quantum circuits for the operators S_1 and S_2, cf. with **Table 1**, for a D2Q9 model define on a 8×8 lattice. The first three qubits are used to encode the position in the x direction and the latter three for the position in the y direction. They are applied to the quantum state only if the velocity register has value 1 or 2 respectively.*

$$\mathcal{B}f_i = -\mathbf{v}_i \cdot \nabla f_i - \frac{f_i - f_i^{eq}}{\tau}, \tag{14}$$

can be expressed in terms of the second-quantization annihilation and generation operators via the relation $\nabla = (\hat{a} - \hat{a}^+)$. Collisions follow the bosonic encoding first proposed by Mezzacapo et al. [42], whereby dissipative effects are represented as a weighted sum of two unitary operators. The scheme is as "Feynmansque" as it can possibly get, since it builds on a one-to-one analogy between LB and the Dirac equation, as first proposed in Ref. [43]. As a result, the formal solution $f_t = e^{\mathcal{B}t} f_0$ can be computed in analogy with quantum mechanics. However, it is subject to a number of additional questions: primarily truncation effects due to the finite number of bosonic excitations and the long-time behavior of non-unitarity errors [44].

On a similar but operationally different line, Sanavio et al. [11] developed a quantum circuit implementing the collision step of a Carleman-Lattice Boltzmann algorithm. Since the number of variables with the Carleman method increases exponentially with the truncation order k, the circuit needs more qubits to embed the relevant information. The collision step requires $k+1$ additional quantum registers, embedding the truncation order and the products between local Boltzmann functions with different velocities, say $f_i(x), f_i(x) f_j(y), f_i(x) f_j(y) f_k(z)$ and so on. With this setting, the collision quantum circuit is genuinely local, as it depends only on the velocity registers and it can be implemented using only a fixed number of two-qubit gates, regardless of the number of lattice sites. This meets the promise of quantum advantage but unfortunately it is inconsistent with the streaming process, which must therefore be directed to a classical machine. This may seem an ideal option for hybrid quantum computing, since streaming is a floating-point free operation. However, accessing data is by no means a cost-free operation and it is indeed known to take a significant fraction of the computer time spent on collisions [45].

5.3 Fully quantum algorithm: collision and streaming

In Ref. [11], the authors managed to generalize the circuit for the streaming step to Carleman variables of higher orders, thus allowing the subsequent application of streaming and collision step on Carleman variables with a single embedding and a single readout of the results after evolution. This results in a fully quantum algorithm. In order to embed different nonlinear terms one needs $2(k+1)$ additional quantum registers, one encoding the information about the degree of non-linearity up to order k, and the other $2k$ to embed position and velocity of the multiproduct functions.

This method was found to meet with an exponential depth of the quantum circuit as a function of the number of qubits. The reason is that the collision operator cannot be written in the form of a sparse and combinatorially block-diagonal matrix, nor in terms of a compact combination of Pauli matrices. Hence, to date, it cannot be quantum compiled.

Possible ways out are currently under investigation: one option is the application of the Carleman linearization method directly to the Navier-Stokes equations. The advantage of the Carleman-Navier-Stokes procedure rests in a much lesser number of Carleman variables, with a corresponding reduction of the circuit depth. The flip side, however, is a more complex structure of the Carleman matrix, which, combined with the Reynolds versus Mach number as a measure of nonlinearity, may impair Carleman convergence. A detailed analysis along this direction is currently underway.

6. Conclusions

In summary, we have presented a survey of the few leading approaches to the quantum simulation of classical fluids. Various obstacles stand in the way of the efficient simulation of fluid flows on quantum computers, especially at high Reynolds numbers, mostly on account of the strong nonlinear effects.

In this chapter, we focused on the Quantum Carleman-Lattice Boltzmann algorithm. Our analysis, informed by prior works [7, 11, 38, 41, 44], indicates that Carleman linearization offers a promising route for modeling low-to-moderate Reynolds number flows, achieving errors on the order of 10^{-3} for Re in the hundreds up to hundred of time steps. However, a significant limitation lies in the exponential depth of the corresponding quantum circuit. Future research should focus on optimizing quantum CLB's resource requirements, potentially through alternative linearization strategies [46] or tailored quantum circuit design. This work highlights the potential of quantum algorithms for fluid simulations, while underscoring the crucial need for addressing their computational demands to unlock real-world computational fluid dynamics applications.

Before closing, it should be observed that while, everybody would expect quantum computers to handle turbulence, the physics of fluids is littered with interesting problems at low-Reynolds, especially in microfluidics, soft matter and biology [37, 47, 48], but also in high-energy physics, typically quark-gluon plasma experiments. Hence, even in case turbulent flows would prove beyond reach to quantum computing, there would nevertheless be plenty of interesting fluid applications in the low-moderate Reynolds regime.

For instance, it could be of great interest to devise a *quantum multi-scale* application, coupling quantum algorithms for biomolecules in a water solvent described by a quantum algorithm for low-Reynolds fluid flow. Given that low-Reynolds flows are non-local, perhaps the inherent nonlocality of quantum mechanics could prove helpful in representing the classical nonlocality of low-Reynolds flows.

On a more philosophical ground, should the physics of fluids show a case for "classical advantage," there would still be interesting lessons to learn. Indeed, as observed earlier on, while it is true that nature is not classical, it is equally true that nature has a very strong tendency to *become* classical at macroscopic scales/high temperatures. Withstanding such a tendency, or perhaps even taking advantage of it, is indeed the prime struggle of quantum computers. Looking into ways to achieve this very hard goal is not only of practical but also of major foundational interest, since it is likely to shed new light into the still open problem of "classicalization," that is, the emergence of classical behavior out of underlying quantum mechanical microscopic physics. This is only one (outstanding) example of the fact that quantum computers offer the unique chance to ask questions that the founding fathers of quantum mechanics could only formulate as "Gedanken Experiments."

Acknowledgements

The authors have benefited from valuable discussions with many colleagues, particularly P. Coveney, N. Defenu, G. Galli, M. Grossi, B. Huang, W. Itani, A. Mezzacapo, S. Ruffo, A. Solfanelli K. Sreenivasan, R. Steijl and T. Weaving. The authors also acknowledge financial support from the Italian National Centre for HPC, Big Data and Quantum Computing (CN00000013).

Author details

Claudio Sanavio[1]* and Sauro Succi[1,2]

1 Fondazione Itituto Italiano di Tecnologia, Italy

2 Mechanical Engineering Department, University College London, UK

*Address all correspondence to: claudio.sanavio@iit.it

References

[1] Nielson MA, Chuang IL. Quantum Computation and Quantum Information: 10th Anniversary Edition. UK: Cambridge University Press; 2010

[2] Feynman R. Simulating physics with computers. International Journal of Modern Physics. 1982;**6**:467

[3] Deutsch D. Quantum theory, the Church–Turing principle and the universal quantum computer. Proceedings of the Royal Society of London A. 1985;**400**(1818):97-117

[4] Preskill J. Quantum computing in the NISQ era and beyond. Quantum. 2018;**2**:79

[5] IBM Quantum [Internet]. www.ibm.com. Available from: https://www.ibm.com/quantum/technology

[6] Bravyi S, Dial O, Gambetta JM, Gil D, Nazario Z. The future of quantum computing with superconducting qubits. Journal of Applied Physics. 2022; **132**(16):160902

[7] Succi S, Itani W, Sreenvasan K, Steijl R. Quantum computing for fluids: Where do we stand? Europhysics Letters. 2023;**144**(1):10001

[8] O'Malley PJJ et al. Scalable quantum simulation of molecular energies. Physical Review X. 2016;**6**(3): 031007

[9] Vorwerk C, Sheng N, Govoni M, Huang B, Galli G. Quantum embedding theories to siumlate condensed systems on quantum computers. Nature Computational Science. 2022;**2**:424

[10] Bharadwaj SS, Sreenivasan KR. Quantum computation of fluid dynamics. arXiv preprint arXiv: 2007.09147.2020;**15**:1-20

[11] Sanavio C, Succi S. Lattice Boltzmann-Carleman quantum algorithm and circuit for fluid flows at moderate Reynolds number. AVS Quantum Science. 1 June 2024;**6**(2): 023802. DOI: 10.1116/5.0195549

[12] Sprague M, Boldyrev S, Fischer P, Grout R, Gustafson WI, Moser R. Turbulent Flow Simulations at the Exascale: Opportunities and Challenges Workshop, Aug 4–5, 2015, Wahington D.C.: US Department of Energy Office of Scientific and Technical Information

[13] Tennie F, Palmer TN. Quantum computers for weather and climate prediction: The good, the bad, and the noisy. Bulletin of the American Meteorological Society. 2022;**104**(2):E488-500. DOI: 10.1175/BAMS-D-22-0031.1

[14] Joseph I, Shi Y, Porter MD, et al. Quantum computing for fusion energy science applications. Physics of Plasmas. 2023;**30**:010501. DOI: 10.1063/5.0123765

[15] Griffin KP, Jain SS, Flint TJ, Chan WHR. Investigation of quantum algorithms for direct numerical simulation of the Navier-stokes equations. Center for Turbulence Research Annual Research Briefs. 2019:347-363

[16] Harrow AW, Hassidim A, Lloyd S. Quantum algorithm for linear Systems of Equations. Physical Review Letters. 2009;**103**(15):150502

[17] Barenco A, Bennett CH, Cleve R, DiVincenzo DP, Margolus N, Shor P, et al. Elementary gates for quantum computation. Physical Review A. 1995; **52**(5):3457-3467

[18] Berry DW, Ahokas G, Cleve R, Sanders BC. Efficient quantum algorithms for simulating sparse Hamiltonians. Communications in Mathematical Physics. 2007;**270**(2): 359-371

[19] Brassard G, Høyer P, Mosca M, Tapp A. Quantum amplitude amplification and estimation. AMS Contemporary Mathematics. 2002;**305**:53-74

[20] Steijl R. Quantum Algorithms for Fluid Simulations. UK: IntechOpen. 2019. DOI: 10.5722/intechopen.86685

[21] Gaitan F. Finding flows of a Navier-stokes fluid through quantum computing. npj Quantum Information. 2020;**6**:61

[22] Budinski L. Quantum algorithm for the Navier–stokes equations by using the streamfunction-vorticity formulation and the lattice Boltzmann method. International Journal of Quantum Information. 2022;**20**(2):2150039

[23] Kacewiz B. Randomized and quantum algorithms yield a speed-up for initail value problems. arXiv preprint; arXiv:quant-ph/0311148. 2003

[24] Budinski L. Quantum algorithm for the advection–diffusion equation simulated with the lattice Boltzmann method. Quantum Information Processing. 2021;**20**(2):57

[25] Succi S. The Lattice Boltzmann Equation (for Fluids and beyond). UK: Oxford University Press; 2001

[26] Benzi R, Succi S, Vergassola M. The lattice Boltzmann equation: Theory and application. Physics Reports. 1992; **222**(3):145-197

[27] Bernaschi M, Melchionna M. Succi S. Rev. Mod. Phys. 91. 025004 51, Mesoscopic simulations at the physics-chemistry-biology interface. 2019. DOI: 10.1103/RevModPhys.91.025004

[28] Bharadwaj SS, Sreenivasan KR. Hybrid quantum algorithms for flow problems. Proceedings of the National Academy of Sciences. 2020;**120**(49): e2311014120

[29] Shi J, Liu N, Yu Y. Time complexity analysis of quantum algorithms via linear representations for nonlinear ordinary and partial differential equations. Journal of Computational Physics. 2023;**487**:3

[30] Lubasch M, Joo J, Moinier P, Kiffner M, Jaksch D. Variational quantum algorithms for nonlinear problems. Physical Review A. 2021; **101**(1):010301(R)

[31] Carleman T. Application of the theory of linear integration equations to nonlinear systems of differential equations. Acta Mathematica. 1932;**59**:63

[32] Succi S, Itani W, Sanavio C, Sreenvasan K, Steijl R. Ensemble fluid dynamic simulations on quantum computers, arXiv preprint arXiv: 2304.05915. Computers and Fluids. 2024;**270**:106148

[33] Orus R. Tensor networks for complex quantum systems. Nature Reviews. 2019;**538–550**(1)

[34] Gourianov N, Lubasch M, Dolgov S, et al. A quantum-inspired approach to exploit turbulence structures. Nature Computational Science. 2022;**2**:30-37

[35] Giannakis D, Ourmazd A, Pfeffer P, Schumacher J, Slawinska J. Embedding classical dynamics in a quantum computer. Physical Review A. 2022; **105**(5):052404

[36] Liu JP, Kolden HO, Krovi HK, et al. Efficient quantum algorithm for dissipative nonlinear differential equations. PNAS. 2021;**118**(35): e2026805118

[37] Succi S. The Lattice Boltzmann Equation for Complex States of Flowing Matter. UK: Oxford University Press; 2018

[38] Li X, Yin X, Wiebe N, Chun J, Schenter GK, Cheung M, et al. Potential quantum advantage for simulation of fluid dynamics. arXiv:2303.16550v205:1-30 [quant-ph]

[39] Yepez J. Quantum lattice-gas models for computational fluid dynamcs. Physical Review E. 2001;**63**:046702

[40] Todorova BN, Steijl R. Quantum algorithm for the collisionless Boltzmann equation. Journal of Computational Physics. 2020;**409**:109347

[41] Itani W, Succi S. Analysis of Carleman linearization of lattice Boltzmann. Fluids. 2022;7(1):24

[42] Mezzacapo A, Sanz M, Lamata L, Egusquiza IL, Succi S, Solano E. Quantum simulator for transport phenomena in fluid flows. Scientific Reports. 2015;**5**(1):1-7

[43] Succi S, Benzi R. Lattice Boltzmann equation for quantum mechanics. Physica D: Nonlinear Phenomena. 1993; **69**(3–4):327-332

[44] Itani W, Sreenivasan K, Succi S. Quantum Algorithm for Lattice Boltzmann (QALB) simulation of incompressible fluids with a nonlinear collision term, arXivpreprint arXiv: 2304.05915, (2023). Physics of Fluids. 2024;**36**:017112. DOI: 10.1063/ 5.0176569

[45] Succi S, Amati G, Bernaschi M, Falcucci G, Lauricella M, Montessori A. Towards exascale lattice Boltzmann computing. Computers and Fluids. 2019; **181**:107-115

[46] Joseph I. Koopman–von Neumann approach to quantum simulation of nonlinear classical dynamics. Physical Review Research. 2020;**2**(4): 043102

[47] Stone HA, Stroock AD, Ajdari A. Engineering flows in small devices: Microfluidics toward a lab-on-a-Chip. Annual Review of Fluid Mechanics. 2004;**2004**(36):381-411

[48] Diotallevi F, Biferale L, Chibbaro S, Lamura A, Pontrelli G, Sbragaglia M, et al. Capillary filling using lattice Boltzmann equations: The case of multi-phase flows. The European Physical Journal Special Topics. 2009;**166**:111-116

Chapter 2

Floating-Point Arithmetic with Consistent Rounding on a Quantum Computer

René Steijl

Abstract

Implementation of floating-point arithmetic with consistent rounding is a critical component of many quantum algorithms. Quantum circuit implementations for squaring and division serve as examples here. This work was motivated by ongoing work in developing quantum algorithms for scientific and engineering computing applications, where this type of arithmetic often forms part of the algorithm. A key feature of the work is the use of a reduced-precision floating-point representation of real data specifically designed for near-term future quantum computing hardware with a limited number of qubits (e.g., less than 100) and with an increased level of fault tolerance as compared to current quantum computing hardware. The quantum circuit implementations of the squaring of a floating-point number and the division of two floating-point numbers are detailed here, highlighting similarities in the quantum circuit implementation for the logical steps required for rounding-to-nearest in line with the IEEE 754 standard for the two arithmetic operations. This similarity is an important feature regarding future work where an automated generation of this type of quantum circuit from a set of standard modules and circuit templates is employed.

Keywords: quantum circuit model, quantum arithmetic, conservation laws, floating-point representation, consistent rounding

1. Introduction

In quantum computing [1] research, significant progress is being made in developing quantum information processing hardware. This has so far created prototype quantum computers in research centers, as well as small-scale quantum computers accessible through the cloud to a wider audience. The current machines belong to the Noisy Intermediate Scale Quantum (NISQ)-computing era [2], characterized by the presence of a small number of qubits (typically less than 100) with significant sensitivity to noise and quantum decoherence and very limited quantum error correction. The ongoing quantum hardware development efforts create quantum computers with an ever increasing number of qubits and increasing levels of fault tolerance.

The context of the present work is the ongoing development effort in creating quantum algorithms for computational science and engineering applications implemented using the quantum circuit model on future, more fault-tolerant quantum computers. In this type of quantum algorithm, there is often a need to perform arithmetic operations on real (non-integer) numbers within the quantum circuit-based implementation. In the quantum computing (QC) literature, there is a significant body of work on quantum circuit implementation for integer arithmetic [3–5], which can be re-used in fixed-point representations of real numbers. However, from the history of computing on classical hardware, it is clear that for most applications, a floating-point representation of data is more effective. Once floating-point numerical computation on classical computers became commonplace, the industry standard IEEE 754 was introduced [6]. A key feature of the IEEE standard is that it requires correctly rounded operations, including the correctly rounded arithmetic operations considered here. Typically, rounding to the nearest floating-pointing number available in the destination (output register) is used. A similar standard for floating-point representations on a quantum computer does not yet exist, but is desirable [7].

Recently, a small number of works have focused on quantum algorithms with a floating-point type representation of data. These works include research on floating-point arithmetic [8–13] as well as research on data encoding and representation of quantum images [14, 15].

The present work discusses quantum circuit implementations for two example calculations typically found in computational science and engineering applications: evaluating the square of floating-point numbers (e.g., as part of kinetic energy evaluation) and the division of two floating-point numbers.

The quantum algorithms of interest here often involve the discretization and solution of partial differential equations (PDEs) that govern the conservation of physical quantities such as mass, momentum and energy. Therefore, it is important to have a consistent rounding in the arithmetic operations to avoid compromising the conservation properties of the discretization and solution methods. Rounding to the nearest floating-point number in the output register of an arithmetic operation is the preferred rounding mode in the context of computational science and engineering applications.

The key contributions of this work can be summarized as follows:

- Demonstration of a quantum circuit implementation of floating-point squaring with detailed analysis of how rounding-to-nearest can be achieved;

- Demonstration and analysis of quantum circuit implementation of floating-point division with rounding-to-nearest;

- Detailed discussion of the modular structure and similarities of the quantum circuit implementation of rounding-to-nearest for both arithmetic operations;

- Discussion of prospects for a wider range of quantum floating-point arithmetic.

The rest of this work is organized as follows. Following the review of related work in Section 2, Section 3 briefly reviews and summarizes key QC principles. Then, the floating-point representation used in the quantum circuit implementations is defined in Section 4. Section 5 discusses the quantum circuit implementation of the squaring of a floating-point number. Quantum circuit implementations for floating-point division are detailed in Section 6. Finally, Section 7 summarizes key findings and directions for future work.

2. Brief review of related work

An early work related to quantum circuits for floating-point arithmetic involved the complexity analysis of floating-point addition and multiplication by Häner and co-workers [8], which showed the significant challenges involved in terms of circuit complexity for IEEE-754 type precision. This motivated the present author to introduce floating-point formats for quantum algorithms with a reduced precision [12, 16, 17], such that the required number of qubits and the circuit complexity remain within limits of near-future quantum computers with an increased level of fault tolerance as compared to NISQ-era hardware.

Optimized circuits for floating-point addition and multiplication were detailed by Gayathri et al. [9] with a particular focus on minimizing the number of T-gates [1] in the implementation. The T-count optimized implementation for single-precision floating-point division was also investigated [10]. An efficient quantum circuit implementation for taking the square root of a floating-point number was detailed more recently [11]. In that work, the classical Babylonian algorithm was implemented using circuits for integer division.

Although the previously referenced works detailed key steps in quantum circuit implementations for floating-point arithmetic, the aspect of rounding was not addressed in detail. This formed the main motivation for the present work.

Key challenges in developing quantum circuit implementations for a divider are the reversibility requirement and fault tolerance, as analyzed by Babu and Mia [18]. The fault-tolerance aspect of a quantum circuit implementation of a floating-point divider was recently addressed by Yuan and their colleagues [13]. Implementations of dividers based on the long division approach perform a series of controlled subtractions for which a comparison operation between two integers encoded by qubit strings is required. Efficient quantum circuit implementations of such "comparator" circuits were published by Xia et al. [19]. Highlighting further recent research work related to quantum floating-point arithmetic, the work by Orts et al. [20] shows efficient circuit implementations for detecting the leading zero in a bit string, and operation typically required in subtraction, division, and multiplication operations.

3. Quantum computing essentials

In this section, some key concepts of quantum computing are reviewed. A full treatment can be found in many textbooks, for example, [1]. In the present work, the quantum circuit model is used to represent quantum algorithms, where a series of quantum gate operations act on a quantum state defined by a qubit register with n qubits in a coherent state. **Figure 1** shows two example quantum circuits with three qubits (a, b, and c) with three successive quantum gate operations. Dirac's bra-ket notation is used so that the quantum state defined by a single qubit can be defined as $|q\rangle = a|0\rangle + b|1\rangle$, such that $|a|^2$ is the probability of finding the qubit in state $|0\rangle$ when

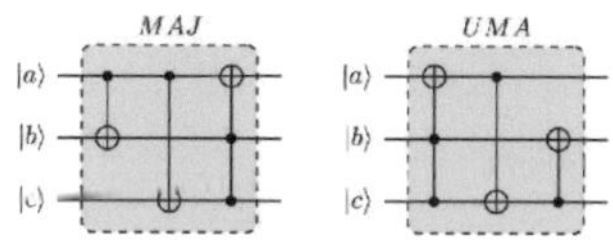

Figure 1.
Definition of MAJ and UMA operations consisting of three gate operations, as used in Cuccaro-type adders.

quantum measurement is performed, and $|b|^2$ is the probability of finding the qubit in state $|1\rangle$. Since a quantum measurement is guaranteed to return one of the two states, it follows that $|a|^2 + |b|^2 = 1$. For a quantum state defined by n qubits in a coherent state, we can write $|q_{n-1}|q_{n-2}| \dots |q_1|q0\rangle = \sum_{i=1}^{2^n-1} C_i$, where the 2^n complex amplitudes C_i $(i \in [0, 2^n - 1])$ define the state. Often, the indices are expanded in binary representation to define the 2^n computational basis states.

In quantum circuit diagrams used in this work, each of the qubits is represented by a horizontal solid line, with a vertical arrangement of the multiple qubits in the register. Quantum gates either work on a single qubit ("single-qubit gates") or work on multiple qubits ("multi-qubit gates"). The quantum computation progresses step-by-step in the horizontal direction, starting from the initial qubit register state defined on the left-hand side of the quantum circuit diagram.

The two example quantum circuits shown in **Figure 1** are the building blocks of quantum arithmetic circuits to be discussed later. On the left-hand side of **Figure 1**, first a controlled X gate is applied. This two-qubit gate applies the X-gate (the quantum computing equivalent of *NOT* in classical circuits) to qubit $|b\rangle$ conditional on qubit $|a\rangle = |1\rangle$. Applying X changes $|1\rangle$ to $|0\rangle$ or $|0\rangle$ to $|1\rangle$. Then, a second controlled-X is applied, now with qubit $|c\rangle$ as the target qubit. Finally, a Toffoli three-qubit gate is applied. This Toffoli gate applies X to the target qubit ($|a\rangle$ here) conditionally on the state of two control qubits. In the rest of this work, "multiply-controlled" versions of the Toffoli gate are used in the quantum circuit diagrams with more than two control qubits.

It should be noted that on actual quantum computing hardware, only a small hardware-specific set of gate operations ("native gates") are implemented. The quantum circuit diagrams used in the present work therefore define algorithms at a higher level of abstraction, and in case of hardware realization, a further translation step toward native gates would be required.

Since a quantum circuit is composed of unitary operations [1], every quantum circuit has an inverse. The inverse of a quantum circuit can simply be composed of the inverse of all its constituent quantum gates, applied in reverse order. Here, it can be noted that the *UMA* circuit shown on the right-hand side of **Figure 1** is not the inverse of the operations from the *MAJ* operation, since the final controlled-X in *UMA* is different from the first controlled-X in the *MAJ* operator.

In the present work, the concept of applying the inverse of an operation defined by a set of quantum gate operations is typically used as *uncomputation*, aimed at returning qubit states to their original state. This is typically used to return qubits allocated to create workspace to their initial state so that these can be re-used in subsequent parts of the circuit.

The quantum circuit implementation of the floating-point dividers considered here is based on modular quantum adders originally introduced by Cuccaro [4]. **Figure 2** shows the example of a 3-qubit full adder, such that an integer "a" defined in 3-bit binary representation as $a2a1a0$ (with $a2$ and $a0$ representing the most significant bit and least-significant bit in the binary representation, respectively) is added to integer "b" (represented as $b2b1b0$ in binary representation). The Cuccaro full adder stores the result in the "b" input. For a full adder, an additional bit is therefore needed for the "b" register to ensure that the output can be represented. The qubit registers $|a2|a1|a0\rangle$ and $|b3|b2|b1|b0\rangle$ correspond to the corresponding bits of "a" and "b." Their initial state on the left-hand side of **Figure 2** defines the intended addition. Upon completion, the quantum state on the right-hand side of the circuit represents the output.

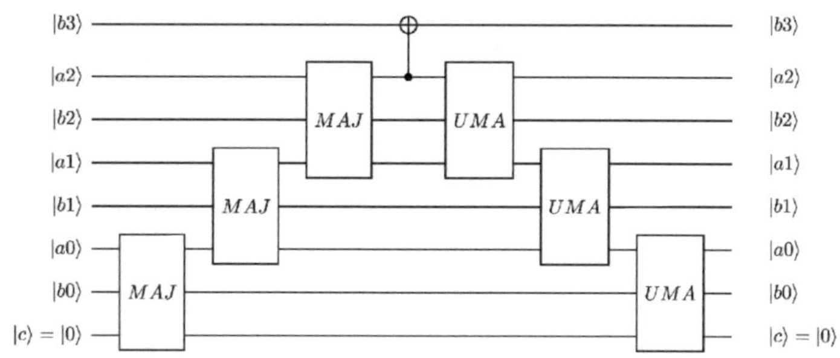

Figure 2.
Illustration of Cuccaro-type full addition. Three-bit integer defined by $|a2|a1|a0\rangle$ *is added to integer defined by* $|b3|b2|b1|b0\rangle$ *with the result stored in* $|b3|b2|b1|b0\rangle$.

The changed state of qubits $|b3|b2|b1|b0\rangle$ relative to that at initialization represents the computational work performed. In later sections, modulo adders will also be used. For the Cuccaro-type adders used here, the underlying idea is that modulo adder can be obtained from the full adder by removing the controlled-X operation in between the top *MAJ* and *UMA* blocks. The second modification involves not using the additional qubit on the "b" register, since by definition it is not needed for a modulo addition. The *MAJ* and *UMA* blocks act on three qubits and were previously defined in **Figure 1**. Furthermore, the quantum circuit implementation includes a "carry" qubit $|c\rangle$, initialized as $|0\rangle$ in the applications considered here. A 3-qubit modulo adder can be obtained from the full-adder circuit shown in **Figure 2** by removing the qubit $|b3\rangle$ and the controlled X operation between the *MAJ* and *UMA* blocks at the top of the circuit. Also, using the concept of 2's complement representation of integer numbers, modulo adder can be used to perform subtraction operations.

4. Floating-point representation in quantum circuits

In previous work, the author introduced a floating-point representation with reduced precision relative to the IEEE-754 single-precision format [12]. A key motivation was the capability to represent a much wider range of real numbers as compared to a fixed-point representation when using the same number of qubits. In this floating-point format, a base 2 representation is used, so that a non-zero, signed real number y is represented in normalized form as:

$$y = \pm SIG \times 2^{EXP}, \quad \text{where} \quad 1 \leq S < 2 \tag{1}$$

where the numbers *SIG* and *EXP* are the mantissa (or significant) and the exponent, respectively. The standard convention for floating-point representations is followed here. *SIG* is defined in a binary fixed-point representation such that the most significant bit is followed by the binary point and the *fractional part* to the right of this point [6]. The mantissa is defined using M bits in a binary representation as,

$$SIG = (m_{M-1}.m_{M-2} \ldots m_1 m_0)_2 \tag{2}$$

Excluding the special cases involving *sub-normal* numbers and *overflow* conditions (to be detailed later), it is assumed that the most significant bit in the binary representation of *SIG* equals 1. Floating-point numbers so defined are termed *normalized* [6]. Eq. (1) shows that normalized numbers cannot exactly represent the real value of zero. To account for this shortcoming and to better represent the smallest numbers in the range of numbers covered by the floating-point representation, *subnormal numbers* are employed. Specifically, for the smallest value of 2^{EXP} in Eq. (1) used in the representation, an additional set of numbers is defined for which the most significant bit in *SIG* equals 0. In the floating-point representation, setting all bits used to define the exponent in binary to 0 (i.e., 00 ... 00) makes the number a sub-normal number. The value for *EXP* used is identical to that used for the smallest of the normalized numbers, that is, with exponents defined as 00 ... 01 in binary representation. Following the IEEE-754 floating-point formats, the exponent bits are also used to identify cases where overflow has occurred in the considered arithmetic operations, that is, where the real value to store does not fit within the range of numbers covered by the floating-point representation. In that case, all exponent bits are set to 1, that is, the exponent is defined as 11 ... 11. More details on using this approach for the IEEE-754 format can be found in standard textbooks, for example, Overton's book [6]. In the quantum circuit implementation, a sign qubit is used which is in state $|0\rangle$ for a positive number. A negative number has this sign qubit in state $|1\rangle$. Following the "hidden-bit" approach used in the IEEE-754 standard [6], the most significant bit is not stored and $M-1$ qubits are used here to store only the fractional part of the mantissa. The exponent is defined using E bits in a binary representation as,

$$EXP = (e_{E-1}e_{E-2}\ldots e_1e_0)_2 - EXP_{bias} \tag{3}$$

where EXP_{bias} is the exponent bias. In this work, 3 qubits are used in the definition of the exponent ($E = 3$), contrasting with the 8 bits used in the IEEE-754 single-precision format. For $E = 3$, sub-normal numbers are defined by the exponent qubit string in state $|000\rangle$. Similarly, the exponent qubit string in state $|111\rangle$ represents overflow conditions. Also, the quantum circuit implementations are detailed for $M = 4$, so that 3 qubits store the fractional part of the mantissa, in contrast to the 23 bits used this in the IEEE 754 single-precision format. This is done for illustration purposes, since for real-world computational engineering application M needs to be significantly larger; for example, it is expected that at least $M = 8$ is required for most applications. For $E = 3$, exponent $(000)_2 = 0$ defines sub-normal numbers, while $(111)_2 = 7$ defines "overflow" conditions. For normalized numbers, this leaves the range 1 to 6, and therefore $EXP_{bias} = 3$ creates a symmetric bias. For this symmetric bias, and assuming $M = 4$, the qubit strings $|011|000\rangle$ and $|011|111\rangle$ define a floating-point number with magnitudes 8/8 and 15/8, respectively (where the sign qubit was omitted for clarity). The smallest normalized number represented by qubit string $|001|000\rangle$ is 8/32. The smallest non-zero sub-normal number equals 1/32 and is represented by the qubit string $|000|000\rangle$. A key aspect of the present format is the use of an asymmetric bias. This enables the range of numbers that can accurately be presented to be tailored to the considered problem. This was demonstrated for quantum circuit implementation of one-dimensional lattice models for fluid dynamics [16, 17]. For the previously considered examples with $M = 4$, an asymmetric bias of $EXP_{bias} = 5$ means that the qubit strings $|011|000\rangle$ and $|011|111\rangle$ define the number 8/32 and 15/32, respectively. Also, for this asymmetric bias, the smallest non-zero sub-normal number represented by qubit string $|000|000\rangle$ takes on the value 1/128.

DOI: http://dx.doi.org/10.5772/intechopen.1005546

5. Quantum circuits for kinetic energy evaluation

This section considers the squaring of floating-point numbers as required, for example, in the evaluation of kinetic energy of a fluid or a solid object. Here, a one-dimensional flow with velocity u is assumed for illustration purposes. Also, to simplify the circuits, no sign qubits are used, with all real data assumed to be positive. The required operation can then be summarized as follows. For an input u defined as a quantum floating-point number with M-qubit representation of the significant ($M - 1$ qubits for fractional part), an integer squaring is performed to create a $2M$−bit temporary result. Based on the exponent of u and the exponent bias used, the next step defines the square in terms of the floating-point format. It is in this second step that rounding aspects are to be considered. Following the definition of the square in floating-point format, the mantissa-squaring step is uncomputed, so that the qubit register at completion of the full quantum circuit for floating-point squaring differs from the initial quantum register state only in terms of the qubits defining kinetic energy (squared number) output.

5.1 Performing mantissa squaring

Integer squaring can be implemented in quantum circuits based on a number of different approaches, e.g., Quantum Fourier-based [5, 12], building on the concept of shift-and-add using controlled additions [17] or optimized logic-based approaches. Since the main focus of this work is the definition of the quantum floating-point output with consistent rounding, this aspect is not analyzed further, and optimized logic-based circuits are used for integer squaring. A 4-bit representation of the mantissa of u is considered here ($M = 4$).

Figure 3 shows the example circuit for the squaring of a 4-bit integer defined in $|a3|a2|a1|a0\rangle$ (with $a3$ the most significant bit in the binary representation), with the 8-qubit integer result defined in $|p7|p6|p5|p4|p3|p2|p1|p0\rangle$. The quantum circuit shown was designed with minimum circuit width in mind, and the creation of five

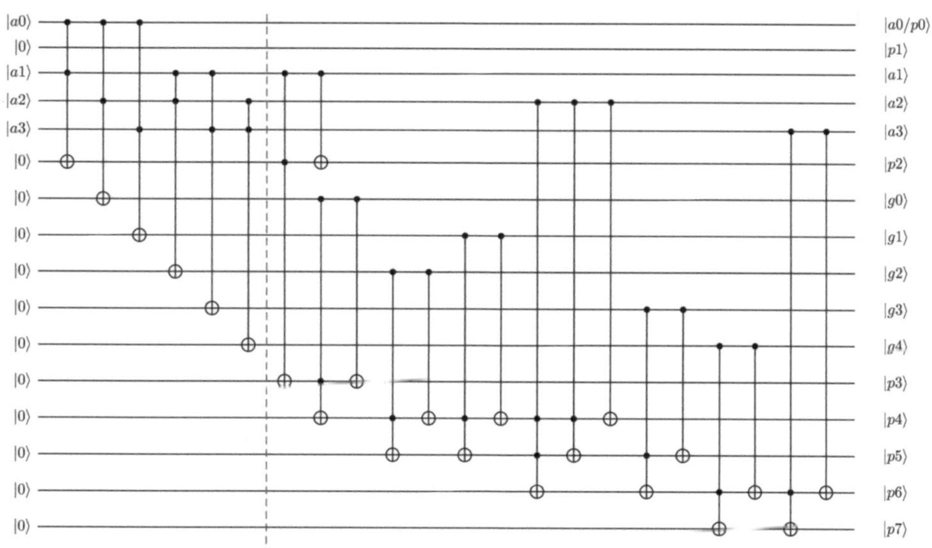

Figure 3.
Quantum circuit for squaring of an integer defined by 4 qubits $|a3|a2|a1|a0\rangle$, *with results in* $|p7|\ldots|p0\rangle$ *such that 5 garbage qubits are created.*

"garbage" qubits $|g0\rangle$ to $|g4\rangle$ was tolerated. Garbage qubits as used here are additional qubits in the quantum circuit used to represent results from intermediate steps in the squaring computation. Assuming that these qubits are initialized in the $|0\rangle$ state at the left-hand side of the circuit, their final state on the right-hand side of the circuit may have changed depending on the particular squaring performed.

Based on the quantum circuit shown in **Figure 3**, a quantum circuit used for mantissa squaring is created that does not create garbage output, but instead uses 5 ancilla qubits that are initialized as $|0\rangle$ and return to $|0\rangle$ at the completion of the squaring. As can be seen, the elimination of garbage qubits comes at the cost of increased circuit depth, that is, an increased number of gate operations. The quantum circuit with 17 qubits is shown in **Figure 4**.

In the quantum circuit shown in **Figure 4**, the 17 qubits represent the following:

$$\begin{array}{lll} |qu3|qu2|qu1|qu0\rangle & : & \text{4 qubits defining 4-bit input 'u'} \\ |r7|r6|r5|r4|r3|r2|r1|r0\rangle & : & \text{8 qubit defining output} \\ |anc4|anc3|anc2|anc1|anc0\rangle & : & \text{5 ancilla qubits} \end{array} \tag{4}$$

Using computational-basis encoding [1], the quantum circuit for squaring can be used as follows. The 17-qubit register is first initialized such that a single complex amplitude is non-zero, that is, it is set to 1. The index of this single non-zero complex amplitude is fully defined by the input integer number, that is, the binary representation of the number defined by $|qu3|qu2|qu1|q0\rangle$. For an integer defined by 4 qubits, the 8 possible outcomes (for a most significant qubit $|qu3\rangle = |1\rangle$) can be summarized as:

$$\begin{array}{llll} u = 8 & C_{in}[1000, 00000000, 00000] & \rightarrow & C_{out}[1000, 01000000, 00000] \\ u = 9 & C_{in}[1001, 00000000, 00000] & \rightarrow & C_{out}[1001, 01010001, 00000] \\ u = 10 & C_{in}[1010, 00000000, 00000] & \rightarrow & C_{out}[1010, 01100100, 00000] \\ u = 11 & C_{in}[1011, 00000000, 00000] & \rightarrow & C_{out}[1011, 01111001, 00000] \\ u = 12 & C_{in}[1100, 00000000, 00000] & \rightarrow & C_{out}[1100, 10010000, 00000] \\ u = 13 & C_{in}[1101, 00000000, 00000] & \rightarrow & C_{out}[1101, 10101001, 00000] \\ u = 14 & C_{in}[1110, 00000000, 00000] & \rightarrow & C_{out}[1110, 11000100, 00000] \\ u = 15 & C_{in}[1111, 00000000, 00000] & \rightarrow & C_{out}[1111, 11100001, 00000] \end{array} \tag{5}$$

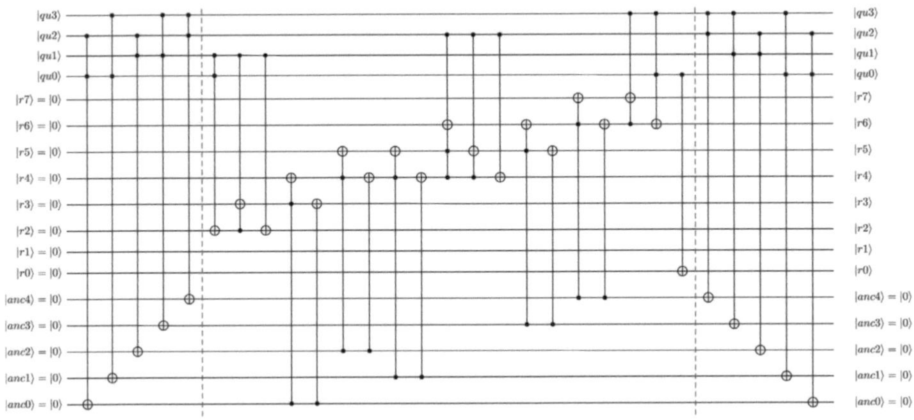

Figure 4.
Quantum circuit used to compute square of the mantissa of u in floating-point squaring.

where $C_{in}[indx_0]$ the complex amplitude in the initial state vector with non-zero magnitude at index $indx_0$ represented in binary representation and the left-hand side above. At the completion of the quantum circuit, a single complex amplitude $C_{out}[indx_1]$ will have non-zero amplitude at index $indx_1$ defined in binary representation on the right-hand side of the listed results.

5.2 Computing u^2 with rounding-to-nearest

For the quantum circuit implementation of squaring u, now defined as a floating-point number with 4 mantissa qubits and 3 exponents qubits, the following qubits are used:

$$\begin{aligned}
&|eu2|eu1|eu0\rangle &&: \text{3 qubits defining exponent of `u'}\\
&|qu3|qu2|qu1|qu0\rangle &&: \text{4 qubits defining mantissa of `u'}\\
&|r7|r6|r5|r4|r3|r2|r1|r0\rangle &&: \text{8 qubit defining output}\\
&|anc4|anc3|anc2|anc1|anc0\rangle &&: \text{5 ancilla qubits}\\
&|esq2|esq1|esq0\rangle &&: \text{3 qubits defining exponent of output}\\
&|msq2|msq1|msq0\rangle &&: \text{3 qubits defining mantissa of output}\\
& && \left(\text{where `hidden-qubit' approach is used}\right)
\end{aligned} \tag{6}$$

For the considered floating-point squaring, the quantum circuit shown previously in **Figure 4** is first applied on the subset of 17 qubits defined previously. This defines the squaring output in the 8 qubits $|r7|r6|\ldots|r0\rangle$ with $|r7\rangle$ defining the most significant bit. This 8-qubit register along with the exponent of the input u is then used to define the output u^2 in terms of the quantum floating-point format. Assuming $EXP_{bias} = 8$, a quantum circuit implementation of the required logical operations is shown in **Figure 5**. Here, the dash slices separate sub-circuits intended for different input exponents.

For the used exponent bias, cases with $|eu2|eu1|eu0\rangle = |000\rangle$ (i.e., sub-normal input u) and $|eu2|eu1|eu0\rangle = |001\rangle$ are guaranteed to create u^2 truncated to 0. This can be seen from the fact that u defined by $|001|111\rangle$ corresponds to a value of $(15/8) \times 2^{1-8} = 15/1024$ and therefore, u^2 is smaller than half the smallest non-zero sub-normal number defined by $|000|001\rangle$ (or $1/1024$) so that rounding-to-nearest always leads to 0.

The left-hand side of the quantum circuit shown at the top of **Figure 5** deals with cases defined by $|eu2|eu1|eu0\rangle = |010\rangle$. As can be seen, for u inputs leading to $|r7\rangle = |1\rangle$, rounding up can then create the smallest non-zero sub-normal output $|000|001\rangle$. The next block of gate operations defines the required logic for $|eu2|eu1|eu0\rangle = |011\rangle$. As a first step, output mantissa qubits $|msq1|msq0\rangle$ are initialized using the state of $|r7|r6\rangle$. In case rounding-by-truncation would have been used, the work required for this input exponent would be completed. However, for the rounding-to-nearest employed here, cases need to be identified for which rounding up is required, here represented by $|anc1\rangle = |1\rangle$ (i.e., ancilla qubit temporarily set to $|1\rangle$). Following a similar approach, the other blocks of gate operations shown in **Figure 5** account for u inputs with $|eu2|eu1|eu0\rangle = |100\rangle$ as well as $|eu2|eu1|eu0\rangle = |101\rangle$ or $|110\rangle$. In all cases, the output mantissa qubits are initially defined according to the rounding-by-truncation (or rounding down) approach, followed by an identification of cases requiring a subsequent rounding up. The rounding up is implemented using multiply-controlled X gate operations creating an increment to mantissa qubits (using

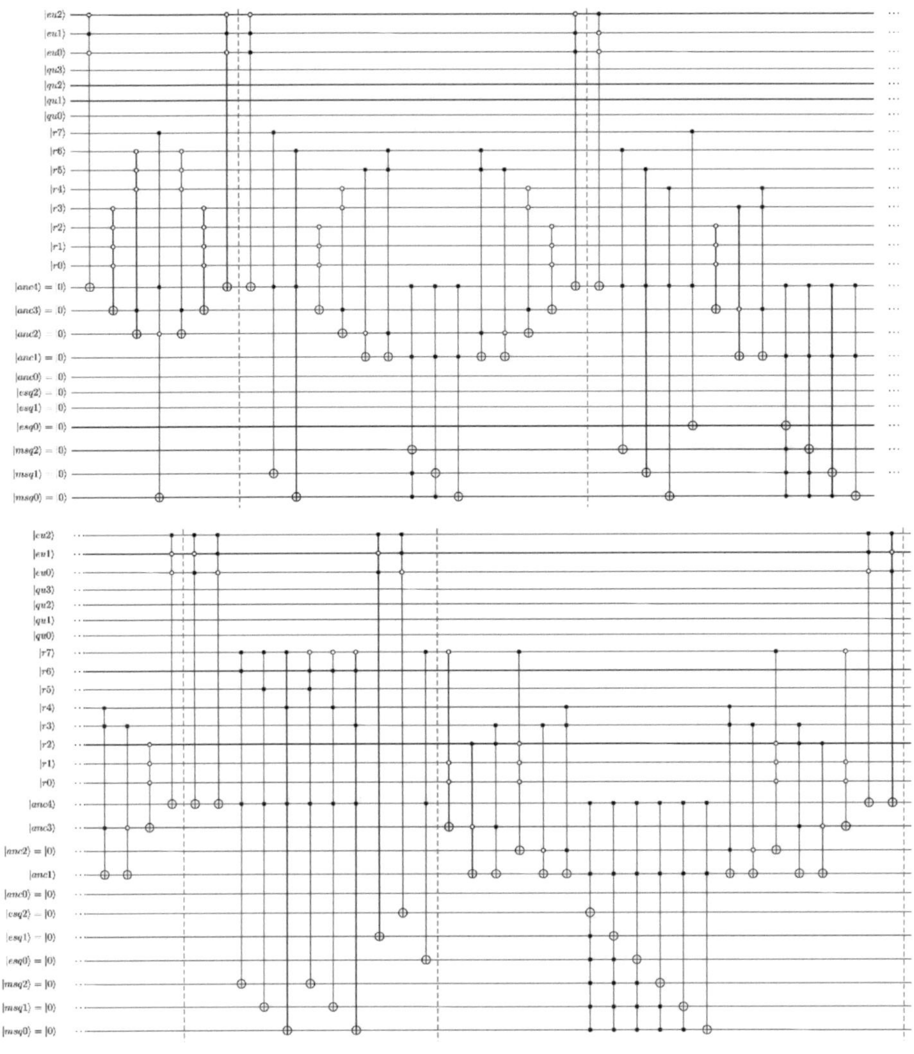

Figure 5.
The circuit shown sets results for u^2 in floating-point format for $M = 4$ using "rounding-to-nearest" ($E = 3$, with $EXP_{bias} = 8$).

hidden-qubit approach) and exponent qubits. For input u with $|eu2|eu1|eu0\rangle = |100\rangle$, the increment operation applied for cases with rounding up can transform the largest sub-normal number $|000|111\rangle$ into the smallest normalized output $|001|000\rangle$. For input cases with $|eu2|eu1|eu0\rangle = |101\rangle$ or $|110\rangle$, the output is guaranteed to be a normalized number and therefore the increment operation performed for rounding up can be seen to include controlled-X gate operations on all the output exponent qubits $|esq2|esq1esq0\rangle$ so that output exponent can be incremented by 1 where required.

Identifying cases that require rounding up is performed as follows:

- If the most significant qubit in $|r7|r6|\ldots|r1|r0\rangle$ that is not yet used to define the initial state of the output mantissa qubits is in state $|1\rangle$, rounding up is needed if any of the further less significant qubits in $|r7|r6|\ldots|r1|r0\rangle$ is in state $|1\rangle$;

- Identifying a "tie." If the most significant qubit in $|r7|r6|...|r1|r0\rangle$ that is not yet used to define the initial state of the output mantissa qubits is in state $|1\rangle$ with all other less significant qubits in state $|0\rangle$, a tie has occurred. In that case, rounding up is applied if the least significant qubit in the initialized output mantissa is in state $|1\rangle$, that is, $|msq0\rangle = |1\rangle$.

The approach used here is consistent with the rounding-to-nearest approach used in the IEEE 754 standard.

6. Quantum circuits for floating-point division

Building on the concepts illustrated in the quantum circuit implementation of floating-point squaring with rounding-to-nearest for the output, the more complex example of floating-point division will be detailed. It is shown here that for the floating-point divider, the quantum circuit implementation of logical steps required to perform rounding-to-nearest for the output displays significant similarity with those for the squaring detailed in Section 5.2.

6.1 Non-restoring and restoring integer dividers

First, quantum circuit implementations for *non-restoring* and *restoring* integer dividers are discussed, to pave the way for the mantissa division required in the floating-point division circuits discussed later. The integer divider considered here performs the division of a 4-bit integer dividend by a 4-bit integer divisor using the concept of long division. Since the aim is to re-use the resulting quantum circuits later on in the quantum floating-point dividers (where it is assumed that both dividend and divisor are normalized numbers), the assumption is made that 4 qubits define the dividend $|q3i|q2i|q1i|q0i\rangle$ with $|q3i\rangle = |1\rangle$, and similarly 4 qubits define the divisor, that is, $|q3d|q2d|q1d|10d\rangle$ with $|q3d\rangle = |1\rangle$.

Before detailing the quantum circuit shown in **Figure 6**, the underlying principle of a quantum circuit implementation based on long division is summarized. To achieve a 4-bit quotient as output of the division by a 4-bit divisor, it follows that the dividend needs to be a 7-bit integer. Starting from the initial 4-bit (input) dividend, a 7-bit workspace is created by padding three 0 bits at the least-significant end (effectively multiplying the integer value by eight). The long division then proceeds as follows. Starting from the 4 most significant bits in the 7-bit workspace, a *comparator* operator (denoted as C) is used to check if the binary value represented by the 4 bits equals or exceeds the divisor value. If so, the most significant qubit defining the quotient will be set to $|1\rangle$. Uncomputing the comparator C (denoted by $U(C)$) restores the 7-bit register to its state before applying C. Then, subtraction by the value of the divisor (denoted by Sub) is applied conditionally on the state of the corresponding quotient qubit. These steps are then repeated a further three times, each time shifting the 4-bit sub-string of the workspace the comparators work on by one bit toward the least significant bit. For the illustrated division for 4-bit dividend and divisors, the 4 repeated steps will fully define the 4-qubit quotient. The final subtraction steps are not needed to define this quotient, as shown in **Figure 6**. The $SWAP$ operations shown in **Figure 6** are there to arrange the qubits in the correct ordering for successive applications of the comparator C, uncomputation of comparator $U(C)$, and the

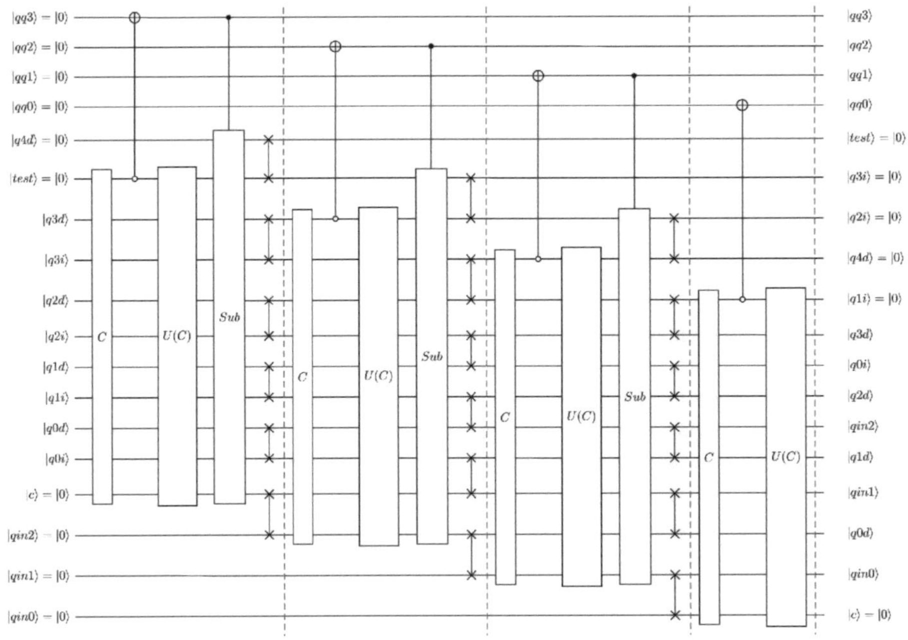

Figure 6.
Quantum circuit for non-restoring integer division based on 7-qubit integer register and 4-qubit divisor. Creates 4-qubit output $|qq3|qq2|qq1|qq0\rangle$.

controlled subtractions. All three operations employ a qubit ordering consistent with Cuccaro-type modulo adders.

For the quantum circuit implementing a non-restoring integer divider defined in **Figure 6**, the qubits are named as follows:

$$\begin{array}{lcl} |qq3|qq2|qq1|qq0\rangle & : & \text{4 qubits defining 4-bit quotient} \\ |q3d|q2d|q1d|q0d\rangle & : & \text{4 qubits defining 4-bit divisor} \\ |q4d\rangle & : & \text{additional qubit used in 2s complement representation} \\ |q3i|q2i|q1i|q0i\rangle & : & \text{4 qubits defining 4-bit dividend} \\ |test\rangle & : & \text{qubit us as 1st comparator output} \\ |c\rangle & : & \text{`carry' qubit in Cuccaro-type arithmetic} \\ |qin2|qin1|qin0\rangle & : & \text{3 additional qubits in 7-qubit workspace} \end{array} \quad (7)$$

The additional qubit $|q4d\rangle$ is required in the temporary conversion of an unsigned 4-bit representation to a 5-bit 2s complement representation used in the conditional subtractions of the divisor value. The subtractions performed by the sub-circuits termed *Sub* are based on Cuccaro-type modulo adders. Subtraction instead of addition is achieved by converting the "a" input integer of the adder (notation follows that in **Figure 2**) into 2s complement representation, and reversing this after the modulo additions is completed. For the controlled subtraction of the divisor defined by 4 qubits, as illustrated here, a 5-qubit modulo adder is employed involving 11 qubits, as shown in **Figure 6** for the blocks termed *Sub*. As discussed in Section 3, Cuccaro-type modulo adders can directly be derived from the Cuccaro-type full adders detailed previously. A 5-qubit modulo adder can be created using 5 successive *MAJ* sub-circuits, followed by 5 *UMA* blocks, using the same structure as illustrated for the 3-qubit full adder in **Figure 2**.

Using the concept of long division, a 4-bit quotient can be obtained by first expanding the 4-bit dividend to a temporary 7-bit workspace defined by 7 qubits as $|q3i|q2i|q1i|q0i|0|0|0\rangle$, that is, 3 qubits initially in state $|0\rangle$ are padded at the back of the original dividend. Using this 7-qubit representation of the dividend guarantees that 4-qubit quotient results with a non-zero most significant bit. As illustrated in **Figure 6**, long division then creates a quotient defined by 4 qubits as $|qq3|qq2|qq1|qq0\rangle$ using the successive application of comparator operations C (as well as the uncomputation of the comparator operator, $U(C)$) and controlled subtractions. Here, a $|0\rangle$ output of the comparator means that the integer value represented by the 4-bit segment of the workspace considered matches or exceeds the integer value of the divisor. In this case, the subtraction is performed and the corresponding output qubit is set to $|1\rangle$. **Figure 7** shows the quantum circuit implementation of the comparator operation used here based on the work of Xia et al. [19].

It should be noted that the quotient defined by the 4 qubits $|qq3|qq2|qq1|qq0\rangle$ can have $|qq3\rangle = |0\rangle$. In the quantum floating-point dividers considered here, this occurrence of $|0\rangle$ for the most significant quotient qubit is to be avoided since, in that case, the least significant output mantissa qubit cannot be defined if the output (quotient) is a normalized number. To prevent this problem, the following changes can be introduced. First, instead of padding 3 qubits in state $|0\rangle$ to the original 4 qubits defining the dividend, now 4 additional qubits are added to create an 8-qubit workspace. As shown in **Figure 8**, the long division then simply involves an additional step (with comparator and uncomputation of comparator, followed by a controlled subtraction) to define a quotient defined by 5 qubits $|qq4|qq3|qq2|qq1|qq0\rangle$. If this integer divider is employed as part of a quantum floating-point divider with 4-qubit mantissa for dividend, divisor, and quotient, then we can use this 5-qubit output as follows. In case $|qq4\rangle = |0\rangle$, the 4 mantissa qubits of the quotient are taken as $|qq3|qq2|qq1|qq0\rangle$. Alternatively, if $|qq4\rangle = |1\rangle$, then $|qq4|qq3|qq2|qq1\rangle$, so that $|qq0\rangle$ will be discarded.

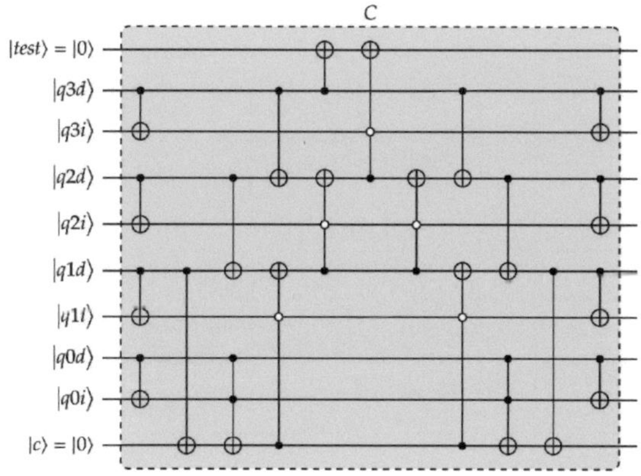

Figure 7.
Example comparator circuit. The result is $|1\rangle$ in cases where the integer defined by $|q3d|q2d|q1d|q0d\rangle$ is greater than the integer represented by $|q3i|q2i|q1i|q0i\rangle$, and $|0\rangle$ in both other cases. For consistency with Cuccaro adders, the ancilla qubit is termed $|c\rangle$.

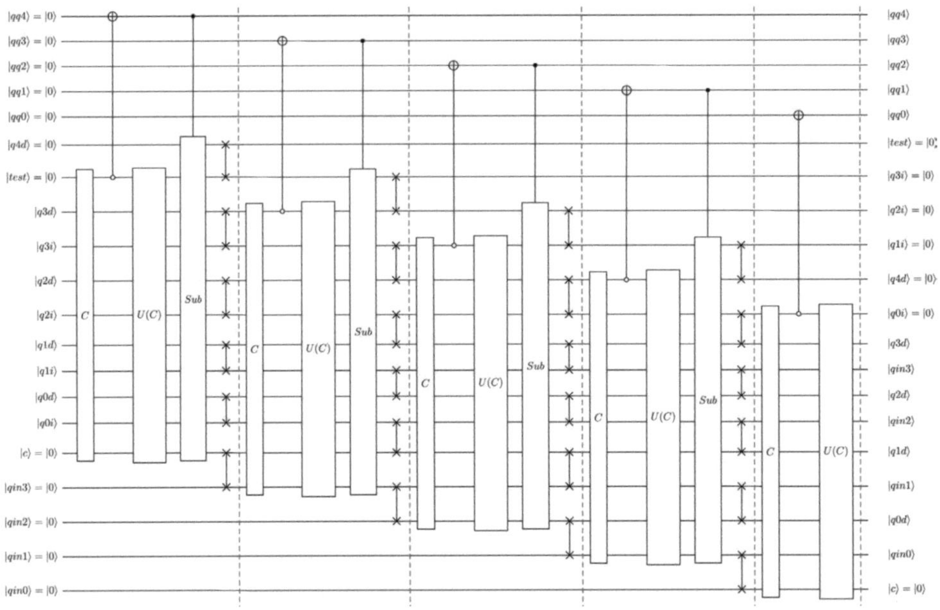

Figure 8.
Quantum circuit for non-restoring integer division based on 8-qubit integer register and 4-qubit divisor. Creates 5-qubit output $|qq4\rangle|qq3\rangle|qq2\rangle|qq1\rangle|qq0\rangle$.

It is important to note that the quantum circuits shown in **Figures 6** and **8** are *non-restoring* division circuits, indicating that upon setting the quotient defined by either $|qq3\rangle|qq2\rangle|qq1\rangle|qq0\rangle$ or $|qq4\rangle|qq3\rangle|qq2\rangle|qq1\rangle|qq0\rangle$, both circuits also change the state of (many) of the other qubits. In applications of integer dividers where such further changes of quantum states are not desirable, a modification of the divider to a *restoring* version is needed. **Figure 9** shows the quantum circuit implementation of a *restoring* integer divider for 4-bit dividend, divisor, and quotient based on a 7-bit workspace.

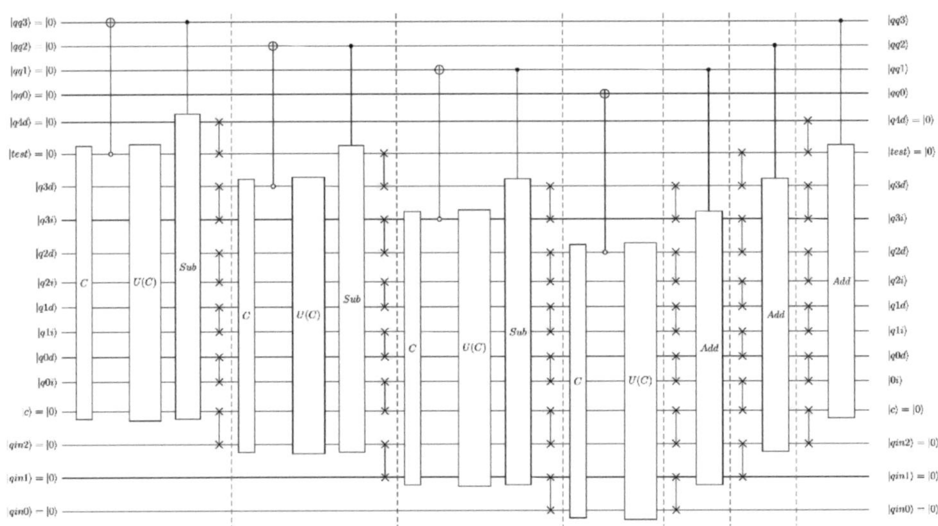

Figure 9.
Quantum circuit restoring integer division based on 7-qubit integer register and 4-qubit divisor. Creates 4-qubit output $|qq3\rangle|qq2\rangle|qq1\rangle|qq0\rangle$.

Comparing the circuit in this figure to the previous *non-restoring* circuit in **Figure 6**, it can be seen that the required changes involve setting up three successive controlled additions to un-compute the effect achieved by the corresponding three controlled subtractions used in defining the quotient qubits. A similar modification to create a restoring version of the non-restoring quantum circuit implementation shown in **Figure 8** would similarly involve a series of 4 controlled additions.

6.2 Extension to floating-point division

We now consider the floating-point division where the dividend, divisor and quotient are all defined in floating-point format with 4 mantissa qubits ($M = 4$), and 3 qubits defining the exponent ($E = 3$).

Based on the integer divider with the 8-qubit workspace, the mantissa division step for a floating-point divider can be created. The use of the 8-qubit workspace guarantees that even if the 5-qubit quotient resulting from long division has its most significant qubit as $|0\rangle$, there are still 4 quotient qubits left to fully define the 4-qubit output mantissa.

In the context of rounding the output of the floating-point division, it is important to consider that at the end of the long-division steps, used to define $|qq4|qq3|qq2|qq1|qq0\rangle$, the 8-bit workspace will be such that the leading 4 qubits $|q3i|q2i|q1i|q0i\rangle$ will be in state $|0|0|0|0\rangle$, while the 4 additional qubits $|qin3|qin2|qin1|qin0\rangle$, which were initialized at $|0|0|0|0\rangle$) now define the *remainder* of the long division.

In the interest of clarity and brevity, divisions leading to quotients defined as normalized floating-point numbers are considered for now. The extension to sub-normal output is discussed later in this section.

As a first step in setting output employing rounding-to-nearest, the output mantissa qubits are initialized using the rounding-by-truncation (or rounding down) approach by considering two different cases. For $|qq4\rangle = |1\rangle$, the 3 output mantissa qubits (defining the fractional part) $|mf2|mf1|mf0\rangle$ are initialized using $|qq3|qq2|qq1\rangle$. For $|qq4\rangle = |0\rangle$, the 3 output mantissa qubits $|mf2|mf1|mf0\rangle$ are initialized using $|qq2|qq1|qq0\rangle$.

To introduce rounding-to-nearest for the quotient, the two different cases also require a different treatment:

- For cases with $|qq4\rangle = |1\rangle$, the state of qubit $|qq0\rangle$ and the state of the remainder of the long-division step defined in qubits $|qin3|qin2|qin1|qin0\rangle$ together define the type of rounding required. Specifically, $|qq0\rangle = |0\rangle$ means that rounding down is to be used, while for $|qq0\rangle = |1\rangle$, the state of qubits $|qin3|qin2|qin1|qin0\rangle$ determines whether rounding up is required, a process that also involves identifying a possible "tie," similar to that described for the floating-point squaring;

- For cases with $|qq4\rangle = |0\rangle$, it is only the state of the "remainder qubits" $|qin3|qin2|qin1|qin0\rangle$ that defines the required type of rounding. In this case, divisions with an even divisor and those with an odd divisor need to be considered separately in terms of identifying the occurrence of a "tie."

Clearly, introducing rounding-to-nearest in a floating-point divider using the 8-qubit workspace restoring integer divider is complicated by the significant differences for the cases with $|qq4\rangle = |1\rangle$ versus cases with $|qq4\rangle = |0\rangle$ as an output of the long division.

To address this problem, an alternative approach is now proposed using a workspace extended by a further qubit. Specifically, starting from a 4-qubit string defining the dividend, 5 additional qubits initiated at state $|0\rangle$ are used at the least-significant end of the register. This effectively multiplies the original number represented by 32, ensures

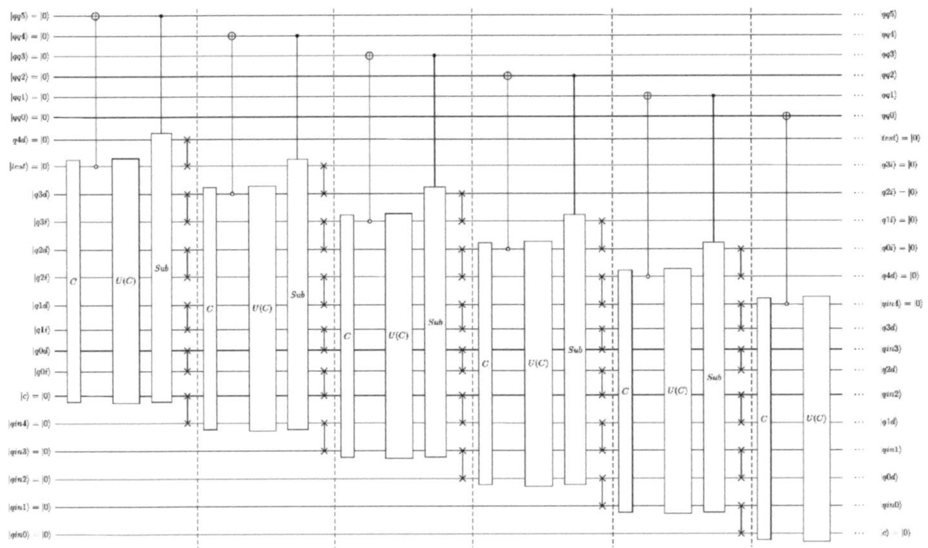

Figure 10.
Quantum circuit for restoring integer division based on 9-qubit integer register and 4-qubit divisor. Creates 6-qubit output $|qq5|qq4|qq3|qq2|qq1|qq0\rangle$. *Part 1.*

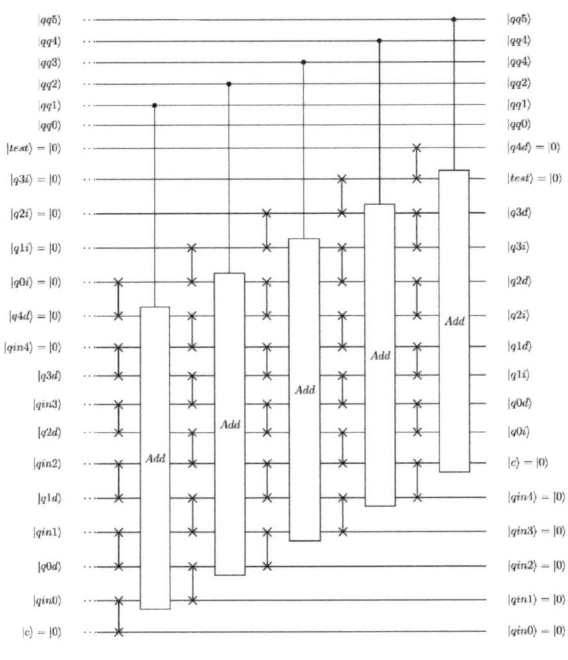

Figure 11.
Quantum circuit for restoring integer division based on 9-qubit integer register and 4-qubit divisor. Creates 6-qubit output $|qq5|qq4|qq3|qq2|qq1|qq0\rangle$. *Part 2.*

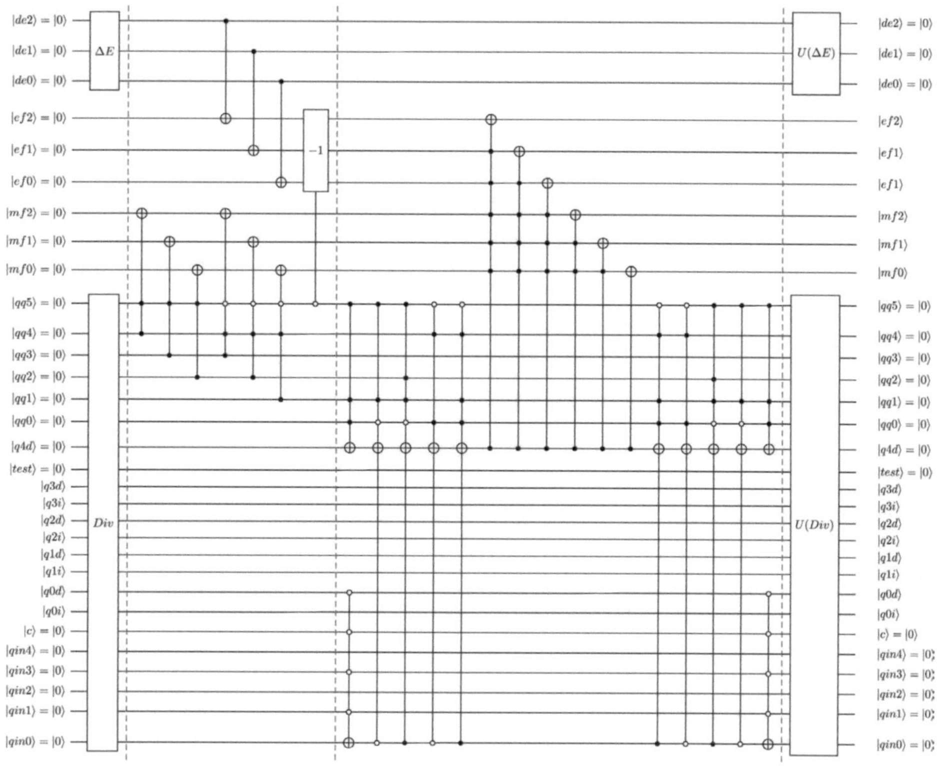

Figure 12.
Quantum circuit implementation of floating-point divider assuming a normalized quotient is created.

that the required 6-qubit quotient is obtained. Of course, this involves creating an additional step in the long division, but the added complexity is accepted here on the basis of the simplified steps for defining and rounding of the output. As will become clear from the following analysis, this approach enables the required logic for rounding-to-nearest of the output to be implemented in quantum circuits following a similar approach introduced for the floating-point squaring in the previous section.

Figure 10 shows the first part of the quantum circuit implementation for the restoring long division used to define the 6-qubit quotient $|qq5|qq4|qq4|qq2|qq1|qq0\rangle$. **Figure 11** shows the part of the quantum circuit implementation that restores all the qubits not defining the quotient output to their original state.

The key idea now is to use the 9-qubit workspace integer division circuits shown in **Figures 10** and **11** for the purpose of the division of the dividend mantissa by the divisor mantissa. In addition to mantissa division, a key role in floating-point division is played by the difference of the dividend of divisor exponents, since this difference largely determines the output exponent (as detailed later).

In the interest of brevity, the quantum circuit implementation considered here includes the following simplifications:

- only positive numbers will be considered here, so there is no need to use qubits defining the sign for dividend, divisor and quotient;

- no qubits are allocated to define the exponent of the dividend and divisor. Instead, three qubits $|de2|de1|de0\rangle$ are allocated initialized at $|0\rangle$. It is assumed that an operator ΔE sets the qubits $|de2|de1|de0\rangle$ to contain the binary representation of the difference between the exponent of the dividend and the exponent of the divisor, with EXP_{bias} subsequently added to this difference.

In **Figure 12**, the quantum circuit implementation of the floating-point divider considered here is shown. The quantum circuit implementation shown assumes that the quotient is a normalized floating-point number. At the end of this section, an extended version without this restriction is presented. Building on the integer divider based on the 9-qubit workspace, the following additional qubits are defined:

$$\begin{aligned} &|de2|de1|de0\rangle &&: \text{ 4 qubits defining the exponent difference} \\ &|ef2|ef1|ef0\rangle &&: \text{ 3 exponent qubits for quotient} \\ &|mf2|mf1|mf0\rangle &&: \text{ 3 qubits for mantissa } \left(\text{`hidden' bit not stored}\right) \end{aligned} \tag{8}$$

In **Figure 12**, *Div* represents the non-restoring part of the integer divider shown in **Figure 10**. On the right-hand side of the circuit, this divider is uncomputed, as indicated by $U(Div)$. Also, at the end of the floating-point division, the operator ΔE is also uncomputed (shown as $U(\Delta E)$) to restore $|de2|de1|de0\rangle$ to $|0|0|0\rangle$.

The quantum floating-point division as shown in **Figure 12** proceeds as follows:

- Operator ΔE defines the difference between dividend and divisor exponents (and adds the exponent bias);

- Long division of the mantissa of dividend and mantissa of divisor, shown as operator *Div* in the quantum circuit;

- Based on the exponent difference set by ΔE and stored in qubits $|de2|de1|de0\rangle$, the exponent of the floating-point output for quotient defined by $|ef2|ef1|ef0\rangle$ is initialized;

- For long divisions resulting in $|qq5\rangle = |1\rangle$, the three qubits $|mf2|mf1|mf0\rangle$ defining the quotient mantissa (using hidden-qubit approach) are set using $|qq4|qq3|qq2\rangle$. For long divisions resulting in $|qq5\rangle = |0\rangle$, the three qubits $|mf2|mf1|mf0\rangle$ are set using $|qq3|qq2|qq1\rangle$;

- For the four qubits defining (part of) the remainder of the long division, $|qin3|qin2|qin1|qin0\rangle$ all in state $|0\rangle$, temporarily set the qubit $|c\rangle = |1\rangle$;

- After completing the long division, the qubit $|test\rangle$ (in state $|0\rangle$) is temporarily in the location of qubit $|q4d\rangle$ during initialization (due to qubit swap operations performed). The qubit $|test\rangle$ will now be set to $|1\rangle$ for cases requiring rounding up;

- For cases with $|qq5\rangle = |1\rangle$, rounding up is always needed for $|qq1|qq0\rangle = |1|1\rangle$ and cases with $|qq1|qq0\rangle = |1|0\rangle$ and $|c\rangle = |0\rangle$. Similar to the floating-point squaring discussed previously in Section 5.2, a tie occurs for $|qq5\rangle = |1\rangle$, $|qq1|qq0\rangle = |1|0\rangle$ and $|c\rangle = |1\rangle$. Then $|mf0\rangle = |1\rangle$ (set by $|qq2\rangle = |1\rangle$) determines cases with rounding up.

- For cases with $|qq5\rangle = |0\rangle$, rounding up is needed for $|qq0\rangle = |1\rangle$ and $|c\rangle = ket0$, while a tie is found for $|qq0\rangle = |1\rangle$ and $|c\rangle = ket1$. The tie-break is decided by the state $|mf0\rangle = |1\rangle$ (set by $|qq1\rangle = |1\rangle$ in this case);

- The following 6 multiply-controlled X operations apply the rounding up by incrementing the mantissa and where needed the exponent;

- The remaining operations perform uncomputation to restore the qubits not defining the quotient output in their original state.

From this outline, it is clear that the fundamental steps involved in defining rounding-to-nearest for the floating-point divider output demonstrate strong similarity with the corresponding steps in the floating-point squaring.

In the final part of this section, an extended quantum circuit implementation of the divider shown in **Figure 12** is considered that can also create quotients defined as sub-normal floating-point numbers. As shown in **Figure 13**, the first change involves making the quantum gate operations used to initialize $|ef2|ef1|ef0\rangle$ and $|mf2|mf1|mf0\rangle$ and the subsequent rounding conditional on the difference of the dividend and divisor exponents. Specifically, in **Figure 13**, the qubit $|q3i\rangle$ (guaranteed to be in state $|0\rangle$ after long division) is temporary set to $|1\rangle$ for cases where the quotient is guaranteed to be a normalized floating-point number. This qubit is then used as (additional) control qubit for the gate operations defining the exponent and mantissa qubits for quotient as well as the subsequent rounding operations. As can be seen, the register $|de2|de1|de0\rangle$ is extended to four qubits by adding $|de3\rangle$, since including the capability to output sub-normal quotients means that the difference in exponents defined by ΔE

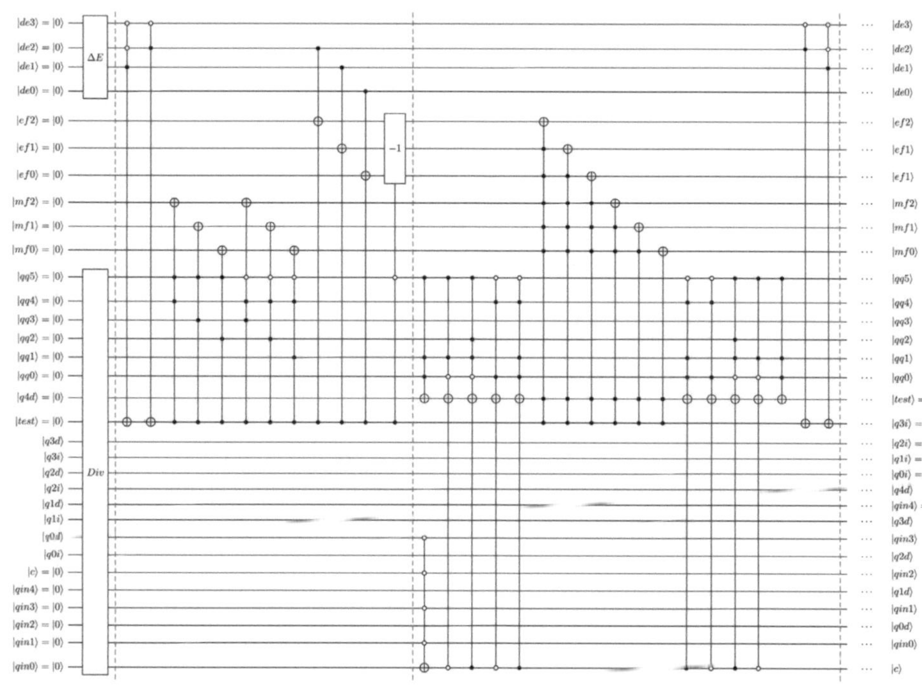

Figure 13.
Quantum circuit implementation of floating-point divider, with quotient defined as either normalized or sub-normal number. Part 1.

now includes negative numbers. Negative numbers are represented using 2's complement formulation, requiring the additional qubit.

The left-hand side of the quantum circuit shown in **Figure 14** considers cases defined by $|de3|de2|de1|de0\rangle = |0|0|0|1\rangle$. For the case a symmetric bias ($EXP_{bias} = 3$ for $E = 3$) is employed, this means that the exponent of the divisor is greater than the exponent of the dividend by a factor of 2. For long divisions leading to results with $|qq5\rangle = |1\rangle$, the quotient is always a normalized number. The largest fraction that needs to be represented is $(15/8)/4 = 15/32$, which can be exactly defined as $|001|111\rangle$ in floating-point format. The smallest fraction that needs to be represented is then $(8/15)/4$ (rounded down to $4/32$ or $|000|100\rangle$ in floating-point format).

The right-hand side of the quantum circuit shown in **Figure 14** considers cases defined by $|de3|de2|de1|de0\rangle = |0|0|0|0\rangle$. The smallest fraction that needs to be represented for $EXP_{bias} = 3$ is then $(8/15)/8$ (rounded down to $2/32$ or $|000|010\rangle$ in floating-point format). Similarly, the largest fraction is $(15/8)/8$ (rounded up to $8/32$ or $|001|000\rangle$ in floating-point format). This shows that using the rounding-to-nearest rounding mode with the required rounding up for the largest fraction is the reason a normalized output results for that case. Rounding-by-truncation creates the subnormal $|000|111\rangle$ for this largest fraction.

In **Figure 15**, the final two possible results for the difference in exponents of dividend and divisor are considered, followed by the uncomputation of the long division and un-computation of ΔE, so that all qubits not used to define quotient output have been restored to their original state. The left-hand side of the quantum circuit shown in **Figure 15** considers cases defined by $|de3|de2|de1|de0\rangle = |1|1|1|1\rangle$ (or '-1' in 2's complement). For the case a symmetric bias is employed, this means that the exponent of the divisor is greater than the exponent of the dividend by a factor of 4. The smallest fraction that needs to be represented is then $(8/15)/16$ (rounded down to $1/32$ or $|000|001\rangle$ in floating-point format). Similarly, the largest fraction is $(15/8)/16$

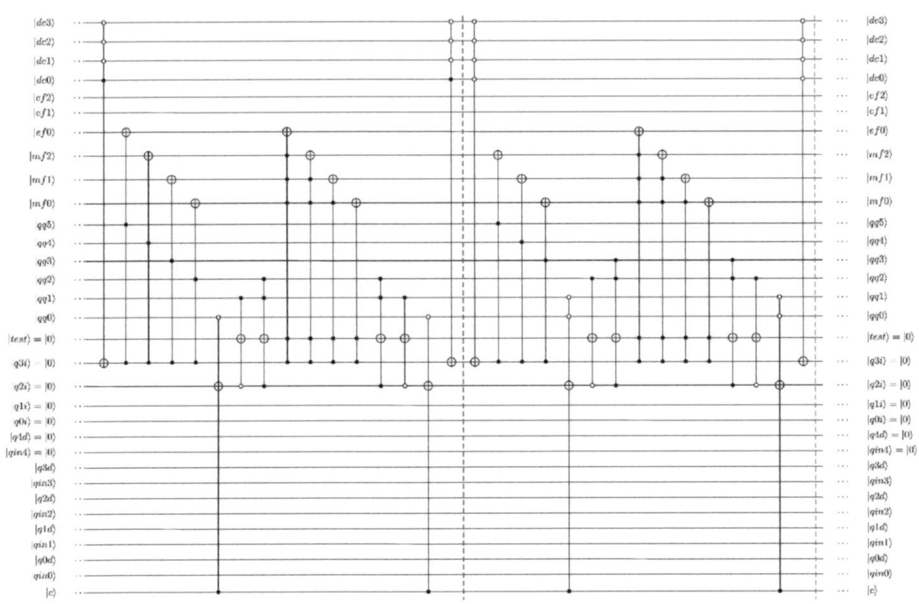

Figure 14.
Quantum circuit implementation of floating-point divider, with quotient defined as either normalized or subnormal number. Part 2.

DOI: http://dx.doi.org/10.5772/intechopen.1005546

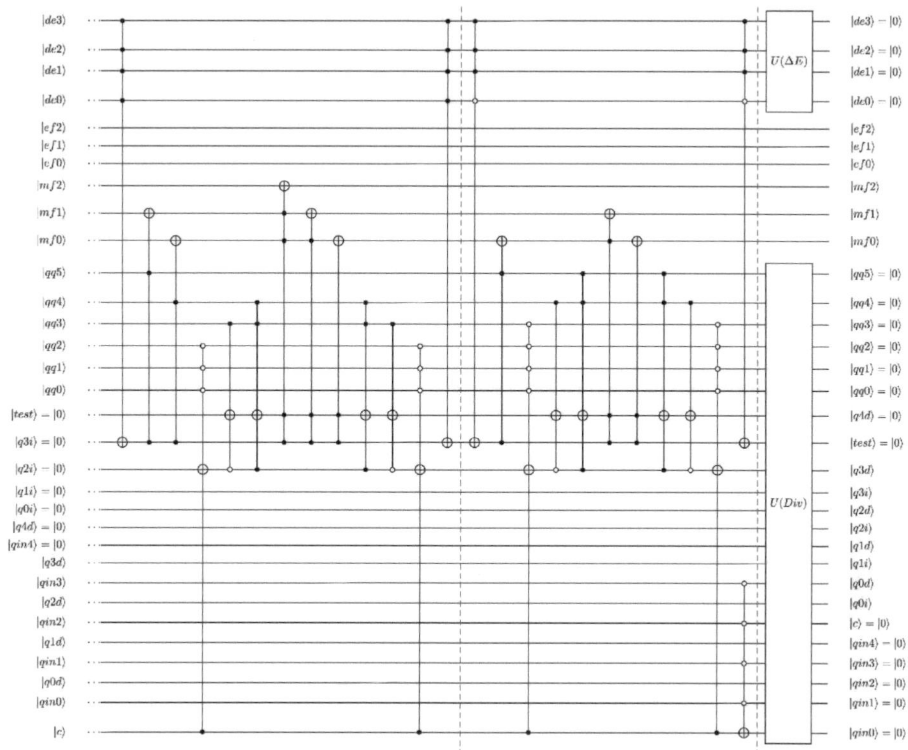

Figure 15.
Quantum circuit implementation of floating-point divider, with quotient defined as either normalized or sub-normal number. Part 3.

(rounded up to $4/32$ or $|000|100\rangle$ in floating-point format). The right-hand side of the quantum circuit shown in **Figure 15** considers cases defined by $|de3|de2|de1|de0\rangle =$ $|1|1|1|0\rangle$ (or $`-2'$ in 2's complement). In that case, the largest fraction to be represented is $(15/8)/32$, with rounding-to-nearest requiring rounding up to $2/32$ or $|000|010\rangle$ in floating-point format. For the smallest fraction $(8/15)/32$, the qubit $|qq5\rangle$ will have state $|0\rangle$, but the rounding-to-nearest requires rounding up to $1/32$ or $|000|001\rangle$ in floating-point format. This means that only the required rounding up needed for the smallest fraction avoids truncation of the output to 0.

Comparing the quantum circuit implementations shown in **Figures 13–15** to the quantum circuits in **Figure 12** clearly shows the significant additional complexity created by dealing with potential sub-normal output. However, the steps involved in performing the rounding-to-nearest of the output remain essentially the same.

7. Conclusions

Two applications of quantum arithmetic involving a reduced-precision floating-point representation were analyzed in detail, with a particular focus on how the floating-point output can be defined using a rounding-to-nearest approach similar to the IEEE-754 standard. The analysis shows that despite the obvious differences in the arithmetic involved in the squaring of a number and the division of two numbers, the quantum circuit implementation of the logic required for rounding-to-nearest for both applications shows significant similarities in the modular designs used here. This is an important finding for future work, in particular when quantum circuit compilation techniques are to be used for the automated definition of quantum arithmetic circuits using a larger number of mantissa and/or exponent qubits than in the examples shown here. In future work, the application to a wider class of arithmetic operations will be considered.

Abbreviations

QC	quantum computing
CFD	computational fluid dynamics
IEEE	Institute of Electrical and Electronics Engineers

Author details

René Steijl
James Watt School of Engineering, University of Glasgow, Glasgow, United Kingdom

*Address all correspondence to: rene.steijl@glasgow.ac.uk

References

[1] Nielsen MA, Chuang IL. Quantum Computation and Quantum Information: 10th Anniversary Edition. 2nd ed. Cambridge: Cambridge University Press; 2010

[2] Preskill J. Quantum computing in the NISQ era and beyond. Quantum. 2018;**2**: 79. DOI: 10.22331/q-2018-08-06-79

[3] Draper TG. Addition on a quantum computer. arXiv.org 2000; 0008033. was cited in original version

[4] Cuccaro SA, Draper TG, Kutin SA, Moulton DP. A new quantum ripple-carry addition circuit. arXiv.org 2004; 0410184. No journal version available

[5] Ruiz-Perez L, Garcia-Escartin JC. Quantum arithmetic with the quantum Fourier transform. Quantum Information Processing. 2017;**16**(6):152. DOI: 10.1007/s11128-017-1603-0

[6] Overton ML. Numerical Computing with IEEE Floating Point Arithmetic. 1st ed. Philadelphia: SIAM; 2001. 97 p

[7] Bhaskar MK, Hadfield S, Papageorgiou A, Petras I. Quantum algorithms and circuits for scientific computing. Quantum Information and Computation. 2016;**16**(3-4):197-236. DOI: 10.5555/3179448.3179450

[8] Haener T, Soeken M, Roetteler M, Svore KM. Quantum circuits for floating-point arithmetic. In: Kari J, Ulidowski I, editors. Reversible Computation. RC 2018, Lecture Notes in Computer Science. Vol. 11106. Cham, Switzerland: Springer; 2018. DOI: 10.1007/978-3-319-99498-7-11

[9] Gayathri SS, Kumar R, Dhanalakshmi S, Kaushik BK. T-count optimized quantum circuit for floating point addition and multiplication. Quantum Information Processing. 2021; **20**:378. DOI: 10.1007/s11128-021-03296-6

[10] Gayathri SS, Kumar R, Dhanalakshmi S, Dooly G, Duraibabu DB. T-count optimized quantum circuit designs for single-precision floating-point division. Electronics. 2021;**10**(6):703. DOI: 10.3390/electronics10060703

[11] Gayathri SS, Kumar R, Haghparast M, Dhanalakshmi S. A novel and efficient square root computation quantum circuit for floating-point standard. International Journal of Theoretical Physics. 2022;**61**:234. DOI: 10.1007/s10773-022-05222-7

[12] Steijl R. Quantum algorithms for nonlinear equations in fluid mechanics. In: Zhao Y, editor. Quantum Computing and Communications. London, UK: IntechOpen; 2022. DOI: 10.5772/intechopen.95023

[13] Yuan S, Gao S, Wen C, Wang Y, Qu H, Wang Y. A novel fault-tolerant quantum divider and its simulation. Quantum Information Processing. 2022; **21**(5):182. DOI: 10.1007/s11128-022-03523-8

[14] Yan F, Chen K, Venegas-Andraca SE, Zhao J. Quantum image rotation by an arbitrary angle. Quantum Information Processing. 2017;**16**:282. DOI: 10.1007/s11128-017-1733-5

[15] Zhang R, Xu M, Lu D. A generalized floating-point quantum representation of 2-D data and their applications. Quantum Information Processing. 2020;**19**:390. DOI: 10.1007/s11128-020-02895-z

[16] Moawad Y, Vanderbauwhede W, Steijl R. Investigating hardware acceleration for simulation of CFD quantum circuits. Frontiers in Mechanical Engineering. 2022;**8**:925637. DOI: 10.3389/fmech.2022.925637

[17] Steijl R. Quantum circuit implementation of multi-dimensional nonlinear lattice models. Applied Sciences. 2023;**13**(1):529. DOI: 10.3390/app13010529

[18] Babu HH, Mia S. Design of a compact reversible fault tolerant division circuit. Microelectronics Journal. 2016; **51**:15-29. DOI: 10.1016/j.mejo.2016.01.003

[19] Xia H, Li H, Zhang H, Liang Y, Xin J. Novel multi-bit quantum comparators and their application in image binarization. Quantum Information Processing. 2019;**18**(7):229. DOI: 10.1007/s11128-019-2334-2

[20] Orts F, Ortega G, Combarro EF, Rua IF, Garzon EM. Optimized quantum leading zero detector circuits. Quantum Information Processing. 2023;**22**:28. DOI: 10.1007/s11128-022-03784-3

Chapter 3

Robust Quantum Algorithms for Early Fault-Tolerant Quantum Computing

Rutuja Kshirsagar

Abstract

Current quantum computing research is divided into two eras: near-term intermediate scaling quantum (NISQ) and fault-tolerant quantum computing (FTQC). There is a significant gap in these both in terms of hardware and algorithms. An important question is "how to transition from NISQ to FTQC"? Some research looks at an intermediate third era of quantum computing, sometimes referred to as "early" fault-tolerant quantum computing (EFTQC). The idea is to establish a trade-off between hardware and algorithmic parameters such that we have limited size circuits, albeit large enough to allow some amount of error-correction at the cost of more number of samples and increased run-time. This also requires adjustment to algorithms for various tasks. One way to do this is to allow the algorithm to tolerate some level of noise in the input. These algorithms are called "robust" quantum algorithms. In this chapter, we will understand what EFTQC means and how we can distinguish between the three eras of quantum computing. Furthermore, we will look at the ideas behind algorithms suitable for EFTQC devices. Lastly, we will look at two examples of robust quantum algorithms for the task of quantum phase estimation.

Keywords: robust quantum algorithms, early fault-tolerant quantum computing, signal processing, Fourier estimation, bridging the gap between NISQ and FTQC

1. Introduction

The field of quantum computing caught attention when it was shown to provide an exponential advantage over classical computations. For example, the Deutsch Jozsa algorithm [1] is a deterministic quantum algorithm which determines whether a given binary function is a constant or a balanced function (outputs 0 for inputs from half of the domain and 1 for the remaining half) and achieves an exponential speed-up over its classical counterparts; Shor's algorithm [2] prime factorizes an integer which also achieves an exponential speed-up over its classical counterparts; Grover's algorithm [3] is an unstructured search algorithm where given an output of a function, the algorithm finds a corresponding unique input with high probability and achieves a polynomial speed-up. Practical implementations of these systems needed development of specific hardware. In particular, a system of error-free qubits. To this end,

IntechOpen

some early developments in the literature of quantum error-correction, such as Shor's 9 qubit code [4], Steane code [5], Knill-Laflamme concatenated quantum codes [6], etc., seemed promising.

These results attracted a lot of people to the field of quantum computing. Tremendous amount of progress has been seen in various aspects of quantum computing from error-correction to algorithm development and from simulations to hardware development. In light of this research, the community has been broadly divided into two categories. First, where the devices are small and noisy. This is usually referred to as the *near-term intermediate scale quantum (NISQ)* era of quantum computing. While the exact scale of devices is not clearly determined, it is assumed that these devices have a couple of 100 qubits at most. Devices in this era of quantum computing are sensitive to noise, are large for simulations due to a large number of degrees of freedom as compared to a classical computer but not sufficiently large to perform error-correction. Furthermore, decoherence errors accumulate in NISQ devices over a period of time, and thus the quantum circuits are constrained by a limited depth.

Quantum algorithms developed for NISQ devices are usually classical-quantum hybrid algorithms, where speed-up is determined by the quantum part of the algorithm and some part of the algorithm includes classical processing. Variational quantum eigensolver (VQE) and quantum approximate optimization algorithm (QAOA) are two popular NISQ algorithms which have various applications. These algorithms require shorter-depth circuits and some amount of error-mitigation to produce accurate results. The VQE algorithm [7] is based on variational methods (approximately finds the ground-state energy), where the quantum computer computes the expectation value of a given system and a classical optimizer improves the guess to determine the ground-state energy of a physical system for a given input ansatz. The QAOA [8], as its name suggests, is an optimization algorithm with a wide range of applications. In most cases, the problem is modeled as a minimization problem to minimize the error. Several variations and applications of VQE and QAOA are studied by the community.

While NISQ era is at one end of the quantum spectrum, at the other end are quantum devices with large number of error-corrected qubits. This is called the *fault-tolerant quantum computing (FTQC)* era. The quantum circuits for algorithms in this regime have deep circuits and support large computations. The threshold theorem [9–11] provides a limit on the maximum physical error allowed in the system below which quantum error-correction schemes can reduce the logical errors exponentially. Some popular fault-tolerant quantum algorithms are quantum phase estimation (QPE) [12], quantum amplitude estimation (QAE) [3, 13], etc.

Currently, there is a large gap between the two eras of quantum computing, namely NISQ and FTQC. We find ourselves at a stage where, we as a quantum community need to bridge the gap between these two eras. The natural question that arises is how? There are several challenges in going from NISQ to FTQC, and therefore there is a new era of quantum computing that is coming up. This is called the *early fault-tolerant quantum computing (EFTQC)* era. Not all research in the literature use the term EFTQC, but the idea is to transition from NISQ to FTQC. This means, that number of physical qubits are still limited (albeit larger than NISQ) at the cost of allowing some amount of (potentially costly) error-correction or more measurements (as compared to FTQC) or other factors. Devices in this new era quantum computing call for a new set of algorithms with some resilience to noise. These algorithms are often called as *Robust quantum algorithms*.

In the next two sections, we will discuss the EFTQC era of quantum computing, robust quantum algorithms and look at some examples of robust quantum algorithms for phase estimation. The fault-tolerant phase estimation algorithm presented in Ref. [12] is discussed in Appendix A.

2. Early fault-tolerant quantum computing

In this section, we will try to understand what exactly determines early fault-tolerant quantum computing. In order to go beyond NISQ, we need larger systems, meaning that the number of physical qubits is large or the circuit has a larger depth. This means that some amount of error-correction is possible. However, the gates (or qubits) are not fully error-corrected. This means that there is still a restriction on the size of the circuit. Given the restriction on the depth of quantum circuits in the NISQ era, the number of measurements required for algorithms to output a "good" outcome is much larger than the number of measurements required by fault-tolerant algorithms run on deeper circuits.

It is therefore crucial to understand the trade-off between parameters of interest to determine how the three eras of quantum computing are separated from each other. One such definition can be found in Ref. [14]. Katabarwa et al. choose physical gate error-rate and physical qubit number as the parameters and define "scalability" as a trade-off between the two. That is, to determine how the error-rates are maintained below (a potentially lower) threshold value while the size of the circuit grows.

The "scalability" model defined in Ref. [14] can be modified to accommodate specific hardware as well as error-correction requirements in order to evaluate the utility of quantum devices. Furthermore, the requirements of algorithms suited for devices in each era of quantum computing may vary. In the next section, we will discuss one particular type of algorithms called *robust* quantum algorithms suited to early fault-tolerant devices.

3. Quantum algorithms for EFTQC

In the early fault-tolerant era of quantum computing, we have circuits deeper that can be deeper than the NISQ circuits with some amount of error-correction and slightly higher measurements as compared to fault-tolerance. Therefore, there arises a need to modify quantum algorithms and make them suitable for EFTQC. The idea is to improve on one or more parameters (reduce the number of logical qubits, reduce the number of gate operations, and increase noise accommodation levels of the algorithm) at the cost of increased number of measurements, complexity as well as classical processing.

Our main focus is one such category of algorithms called *robust quantum algorithms*. Robust quantum algorithms incorporate some amount of noise in the inputs. The idea here is to consider noise as a nuisance parameter to the input function and then run the algorithm. One way to achieve this is minimizing the quantum computation to what is absolutely necessary and offloading other computations to a classical system. However, the classical offloading means the quantum states must be mapped to classical states and back. Given that the quantum systems have more degrees of freedom as compared to their classical counterparts, this back-and-forth mapping of states will create an overhead. This will in turn lead to an increased number of measurements as well as re-initialization of quantum states.

In the next section, we will look at some examples of robust quantum algorithms.

4. Some examples of robust quantum algorithms for quantum phase estimation

In this section, we will look at a nonexhaustive list of robust quantum algorithms for the task of phase estimation. We will particularly focus on quantum phase estimation because the methodologies can be easily extended to develop algorithms for other tasks, such as amplitude estimation, ground-state energy estimation, ground-state property estimation, etc. Appendix A contains a summary of the fault-tolerant quantum phase estimation algorithm [12]. Appendix B contains a summary of the Hadamard test [15], which is used to develop the robust quantum algorithms discussed here.

4.1 Robust phase estimation

We will first look at the Robust Phase Estimation (RPE) algorithm [16] for phase estimation. It is a robust quantum algorithm, meaning it can accommodate for some errors. As long as the errors are below a certain threshold, the algorithm estimates the phase of an input state accurately. The RPE protocol requires application of a unitary U repeatedly, such that given input $|0\rangle$ or $|+\rangle = \frac{|0\rangle + i|1\rangle}{\sqrt{2}}$, the quantum circuit used for measurement encodes the phase with a probability distribution as in Eqs. (1) and (2). Let the experiments be indexed by k, then in the kth generation U is applied N_k times. And, in each run, RPE estimates the rotation of U, $\hat{\theta}_k$. The maximum $\hat{\theta}_k$ is the required phase.

$$Pr_{0,k} = \frac{1 + \cos(N_k\theta)}{2} \tag{1}$$

$$Pr_{+,k} = \frac{1 + \sin(N_k\theta)}{2} \tag{2}$$

The algorithm accommodates some level of noise in the input, which is modeled as additive errors in the probability distribution. That is, Eqs. (1) and (2) are modified as Eqs. (3) and (4) for noise $\delta_0(k)$ when the input is 0 and noise $\delta_+(k)$ when the input is +.

$$Pr_{0,k} = \frac{1 + \cos(N_k\theta)}{2} + \delta_0(k) \tag{3}$$

$$Pr_{+,k} = \frac{1 + \sin(N_k\theta)}{2} + \delta_+(k) \tag{4}$$

Therefore, under the assumption that perfect resources are not required for phase estimation, in the worst-case scenario depolarizing errors or amplitude sampling noise can be re-modeled as additive errors. Furthermore, the set-up allows control over gates (including gate errors) and upon characterization, systematic errors can be undone. This process is robust to state preparation and measurement errors. It may be noted that the algorithm works efficiently as long as the amount of noise is below a certain threshold. The algorithm is Heisenberg-limited, meaning given an accuracy ε the run-time scales as $O(1/\varepsilon)$ in the case that the noise is below the threshold.

The main result of the paper [16] where RPE was introduced can be summarized as follows. Consider a nonadaptive experimental set-up. Given an input 0 or $+$ with noise $\delta_0(k)$ and noise $\delta_+(k)$, respectively, RPE can be performed with probability of success as given in Eqs. (3) and (4). The kth experiment takes time proportional to k and estimates $\hat{\theta}_k \in (-\pi, \pi]$ with a standard deviation $\sigma(\hat{\theta}_k)$. The run-time of the algorithm is $O(1/\sigma(\hat{\theta}_k))$. Moreover, if $|\delta_0(k)|$ and $|\delta_+(k)|$ are less than $1/\sqrt{8}$ for all $k < k^*$ then $\sigma(\hat{\theta}_k) \sim O(1/k^*)$.

4.2 Randomized fourier estimation

In this subsection, we will look at the Randomized Fourier Estimation (RFE) [17] and modified RFE algorithms [18] for phase estimation. The RFE algorithm introduced in Ref. [17] works on a similar principal as the RPE, meaning that it estimates the phase of an input state accurately and efficiently in the presence of noise up to a threshold. While two types of noise models are considered in Ref. [17], another one is considered in Ref. [18]. The circuit used for (lower bounded number of) measurements is a modified Hadamard test (Appendix B) circuit as shown in **Figure 5** (**Figure 1**).

Here, $S = \begin{bmatrix} 1 & 0 \\ 0 & i \end{bmatrix}$ is the phase gate. In the case that $\ell = 0$, the modified Hadamard test (real in this case) produces outcome $c = \pm 1$ with probability given in Eq. (5).

$$Pr(c|k) = \frac{1 + c\cos(k\theta)}{2} \tag{5}$$

In the case that $\ell = 1$, the modified Hadamard test (imaginary in this case) produces outcome $s = \pm 1$ with probability given in Eq. (6).

$$Pr(s|k) = \frac{1 + s\cos(k\theta)}{2} \tag{6}$$

The outcome probabilites in Eqs. (5) and (6) change based on the noise model used to determine the noise in inputs. The analysis uses an analogy to a signal processing task due to similarity. Let us first consider the case where there is noise in the input state. The RFE algorithm can be summarized as follows. Given a maximum number of iterations (samples) M and a maximum depth K (which is analogously the maximum value of the time variable), initialize a Fourier estimate to $0, \hat{f} \leftarrow 0 \in \mathbb{C}^K$. Draw a sample of the time signal as in Eq. (7) for a time variable $k \in \{0, \ldots, K-1\}$ and call the real Hadamard test (output c) as well as the imaginary Hadamard test (output s).

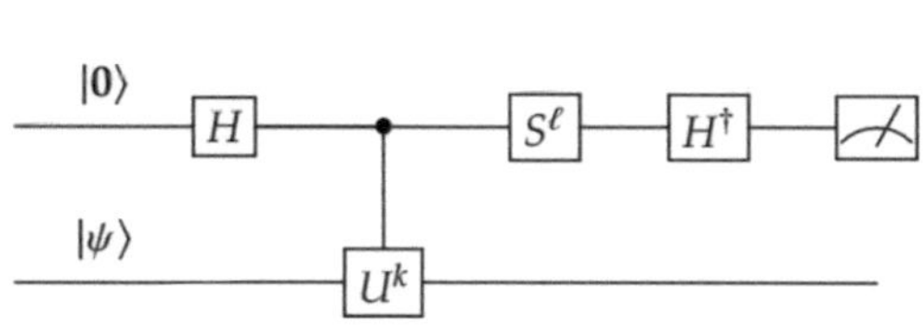

Figure 1.

(figure 1 in [17]) circuit diagram of the Hadamard, where the application of phase gate $S = \begin{bmatrix} 1 & 0 \\ 0 & i \end{bmatrix}$ *is toggled by* $\ell = 0, 1$.

$$g(k) = (Pr(c = 1|k) - Pr(c = -1|k)) + e^{ik\theta}(Pr(s = 1|k) - Pr(s = -1|k)) \tag{7}$$

Update the Fourier estimate M times using Eq. (8). Let j represent the iteration number.

$$\hat{f}_j \leftarrow \frac{\hat{f}_j(c + is)e^{-2\pi ikj/K}}{M} \tag{8}$$

This generates an array of M Fourier estimates. The output phase corresponds to the maximum Fourier estimate. In the noiseless case, the algorithm efficiently estimates the accurate phase with high probability. Given an accuracy ε, RFE estimates the accurate phase with high probability in $O(1/(\varepsilon \ln \varepsilon))$ time.

Now, we will look at how a noisy input affects the algorithm. There are two types of additive noise models used to determine the noise in the input in Ref. [17]. The first model called the bounded adversarial noise (BAN) model represents a worst-case analysis of the algorithm's tolerance to noise. The second model called the Gaussian noise (GN) model represents a more realistic noise model where the additive noise is drawn from a Gaussian distribution. The GN model is a special case of the BAN model. In both these cases, given that the noise is below a certain threshold the number of samples required to accurately estimate the phase is slightly higher than that in the noiseless case. Moreover, the run-time of the algorithm is slightly longer. However, in both the cases given an accuracy ε, RFE estimates the accurate phase with high probability in $O(1/(\varepsilon \ln \varepsilon))$ time.

A third noise model presented in Ref. [18] captures the effect of noise as a combination of exponential decay and random fluctuations in the algorithm. Let us call this the algorithmic noise (AN) model. The analysis requires a modified version of RFE, where the phase gate in modified Hadamard tests is replaced with a modified phase gate, $S(\phi) = \begin{bmatrix} 1 & 0 \\ 0 & e^{i\phi} \end{bmatrix}$. The circuit is presented in **Figure 2**.

Under this noise model, the time signal $g(k)$ in Eq. (7) decays exponentially. This means that identifying the largest Fourier estimate requires a more number of samples. We will not go into the details of the analysis here. Given that the noise is low, the algorithm is Heisenberg-limited. That is, given an accuracy ε, the algorithm runs in $O(1/\varepsilon)$ time. However, if the noise is high, the algorithm runs in $O(1/\varepsilon^2)$ time.

We summarize the outcome probabilities of the real and imaginary modified Hadamard test noiseless cases and all three noise models are discussed in **Table 1**.

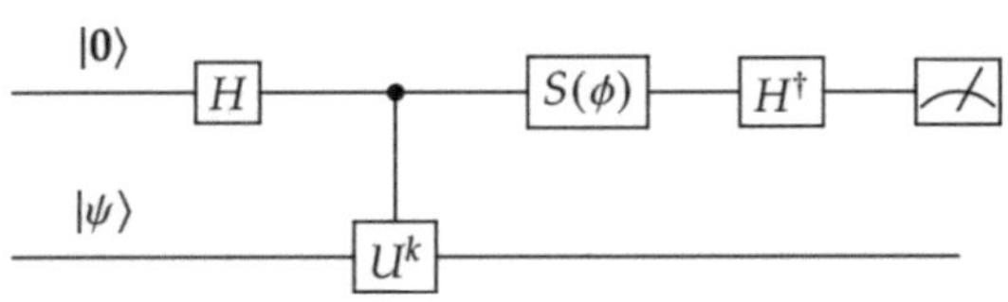

Figure 2.
(figure 1 in [18]) circuit diagram of the modified Hadamard, where the application of modified phase gate $S(\phi) = \begin{bmatrix} 1 & 0 \\ 0 & e^{i\phi} \end{bmatrix}$ *leads to a measurement* $\sigma_\phi = \cos(\phi)\sigma_x - \sin(\phi)\sigma_y$ *of the region of the circuit comprising of the* $S(\phi), H^\dagger$ *gates and the meter; and* σ_x, σ_y *are conventional Pauli operators.*

	Real Hadamard test	Imaginary Hadamard test	Special conditions
Noiseless input	$Pr(c\|k;\theta) = \frac{1+c\cos(k\theta)}{2}$	$Pr(s\|k;\theta) = \frac{1+s\sin(k\theta)}{2}$	None
BAN model	$Pr(c\|k;\theta) = \frac{1+c\left(\cos(k\theta)+\eta_{1,k}\right)}{2}$	$Pr(s\|k;\theta) = \frac{1+s\left(\sin(k\theta)+\eta_{2,k}\right)}{2}$	None
GN model	$Pr(c\|k;\theta) = \frac{1+c\left(\cos(k\theta)+\eta_{1,k}\right)}{2}$	$Pr(s\|k;\theta) = \frac{1+s\left(\sin(k\theta)+\eta_{2,k}\right)}{2}$	$\eta_{1,k}, \eta_{2,k}$ are drawn from a Gaussian distribution.
AN model	$Pr(c\|k,\phi;\theta) = \frac{1+ce^{-\lambda k}\cos(k\theta-\phi)+\eta_{k,\phi}}{2}$	$Pr(s\|k,\phi;\theta) = \frac{1+se^{-\lambda k}\sin(k\theta-\phi)+\eta_{k,\phi}}{2}$	None

Table 1.
Probability of outcomes of real and imaginary Hadamard tests for various models of RFE.

5. Discussion

The Robust Phase Estimation algorithm summarized in Section 4.1 estimates all systematic single-qubit gate errors simultaneously and represents these using additive errors in the measurement probabilities of experiments (see Eqs. (2) and (3)). However, this analysis can be extended further to multi-qubit operations. Furthermore, the additive noise model can be refined to include errors such as due to twirling occurring by repeated gate applications. The algorithm has been tested for robustness in Ref. [19]. We will not go into the analysis of these experiments here. The framework of the algorithm has also been extended to develop other robust quantum algorithms such as robust ground-state energy estimation [20].

Similarly, the Randomized Fourier Estimation algorithm summarized in Section 4.2 provides a framework to analyze quantum algorithms under different noise models. The noise models can be fine-tuned to accommodate a more practical scenario. Furthermore, the framework can be utilized to develop robust quantum algorithms for tasks, such as amplitude estimation, ground-state energy estimation, ground-state property estimation, etc.

6. Conclusions

The quantum community is currently divided between the near-term intermediate scaling quantum (NISQ) era and the fault-tolerant quantum computing (FTQC) era. In the NISQ era, the devices are small and highly sensitive to noise. The small sizes of circuits imply that error-correction is not possible. Algorithms developed for running on NISQ devices are hybrid classical-quantum algorithms. The quantum part of the algorithm requires a shorter-depth circuit, which means a large part of the process must be offloaded to classical systems for processing. Since error-correction is not possible, some error-mitigation techniques might be necessary. Here, we briefly touch upon two NISQ algorithms: VQE and QAOA.

In the FTQC era, the devices are large and resilient to noise. The larger sizes of circuits imply error-correction can be done efficiently. Algorithms developed for running on FTQC devices have deep circuits and support a large number of computations. We have discussed one FTQC algorithm: QPE. It may be noted that the hardware development of FTQC devices is not feasible currently.

In moving from NISQ to FTQC era, the size of circuits must increase to allow at least some amount of error-correction. This will come at some cost in terms of fault-tolerance. Some works in the literature call this the early fault-tolerant quantum computing (EFTQC) era. There are several ways in which the three eras of quantum computing can be distinguished from each other. We discuss one such model: namely the "scalability" model where the scalability is the measure of trade-off between the circuit size and number of samples required to run any algorithm successfully. The scalability model used to distinguish between the three eras of quantum computing can be improved by incorporating several hardware and algorithmic parameters into the model definition.

Devices in the EFTQC era of quantum computing are larger than their NISQ counterparts. This enables some amount of error-correction. This means that error-mitigation and error-correction techniques may be combined to deal with noise. New algorithms may be required to suit EFTQC devices, which could need deeper circuits. These deeper circuits will require a cost in the form of increased measurements and run-times. There are a few different techniques found in the literature to develop these new algorithms. One such technique incorporates some amount of noise in the algorithm input. This class of algorithms is called "robust" quantum algorithms. We discuss two types of robust quantum algorithms for the task of phase estimation: namely RPE and RFE. We do not discuss an exhaustive list of algorithms or development techniques in this work. Yet, there are several ways in which the algorithms can be improved in order to move toward fault-tolerance. All-in-all, we are still far-away from fault-tolerance. However, development of hardware as well as algorithms for this new era of quantum computing, namely EFTQC, might be a step in the direction of slowly achieving fault-tolerance.

Abbreviations

NISQ	Near-term intermediate scaling quantum
FTQC	Fault-tolerant quantum computing
EFTQC	Early fault-tolerant quantum computing
VQE	Variational quantum eigensolver
QAOA	Quantum approximate optimization algorithm
QPE	Quantum phase estimation
QAE	Quantum amplitude estimation
QFT	Quantum Fourier transform
RPE	Robust phase estimation
RFE	Randomized Fourier estimation
BAN	Bounded adversarial noise
GN	Gaussian noise
AN	Algorithmic noise

Appendix A: Kitaev's phase estimation algorithm

Here, we will look at the (fault-tolerant) QPE algorithm. Given a unitary matrix U and its eigenvector $|\psi\rangle$, QPE [15] tries to estimate the angle θ_ψ by estimating $|\psi\rangle$ where $U|\psi\rangle = e^{i\theta_\psi}|\psi\rangle$. Given an additive error ε, the algorithm needs $O(\log(1/\varepsilon))$ qubits and $O(1/\varepsilon)$ controlled-U operations. An important question to answer is what kind of information can we get by running QPE. Let us first look at the single (ancilla) qubit version of QPE. Consider the circuit in **Figure 3**, which is applied to implement (single-qubit) QPE.

Given an input $|0\rangle \otimes |\psi\rangle$, applying the Hadamard gate to $|0\rangle$ yields

$$\frac{1}{\sqrt{2}}(|0\rangle + |1\rangle) \otimes |\psi\rangle = \frac{1}{\sqrt{2}}(|0\rangle \otimes |\psi\rangle + |1\rangle \otimes |\psi\rangle). \tag{9}$$

Then applying controlled-U to $|\psi\rangle$ yields

$$\frac{1}{\sqrt{2}}\left(|0\rangle \otimes |\psi\rangle + e^{i\theta_\psi}|1\rangle \otimes |\psi\rangle\right). \tag{10}$$

Finally applying the Hadamard gate to the first qubit yields:

$$\frac{1}{\sqrt{2}}\left(\left(1 + e^{i\theta_\psi}\right)|0\rangle \otimes |\psi\rangle + \left(1 - e^{i\theta_\psi}\right)|1\rangle \otimes |\psi\rangle\right). \tag{11}$$

The probability of measuring the first qubit as $|0\rangle$ is

$$\left|\frac{1}{2}\left(1 + e^{i\theta_\psi}\right)\right|^2 \tag{12}$$

and that of measuring the first qubit as $|1\rangle$ is

$$\left|\frac{1}{2}\left(1 - e^{i\theta_\psi}\right)\right|^2. \tag{13}$$

The algorithm can be extended to a multiple (ancilla) qubit system. This is the QPE algorithm presented in Ref. [12]. Given an input $|0\rangle^{\otimes n} \otimes |\psi\rangle$, consider the quantum circuit in **Figure 4**.

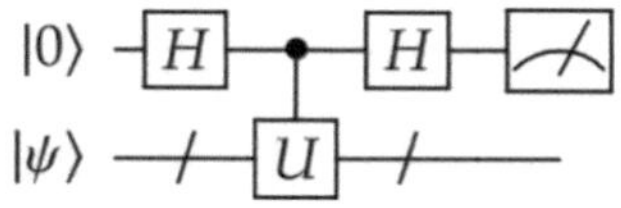

Figure 3.
Circuit implemented for (single-qubit) QPE.

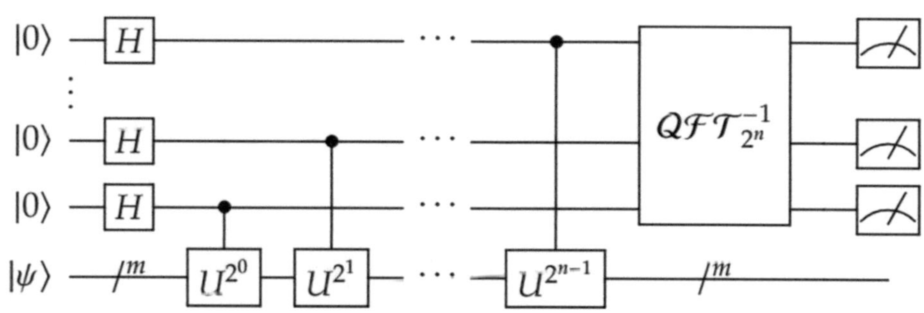

Figure 4.
Circuit implemented for (multiple qubit) QPE.

Applying the Hadamard gate to $|0\rangle^{\otimes n}$ yields

$$\frac{1}{2^{n/2}}(|0\rangle + |1\rangle)^{\otimes n} \otimes |\psi\rangle = \frac{1}{2^{n/2}} \sum_{j=0}^{2^n-1} |j\rangle|\psi\rangle. \tag{14}$$

Then applying controlled-U^j to $|j\rangle|\psi\rangle$ yields

$$\frac{1}{\sqrt{2^n}} e^{2\pi i j\theta} |j\rangle|\psi\rangle. \tag{15}$$

Finally applying the inverse quantum Fourier transform yields

$$\frac{1}{2^{n/2}} \sum_{k=0}^{2^n-1} e^{2\pi i\theta k} \left(\frac{1}{2^{n/2}} \sum_{x=0}^{2^n-1} e^{-2\pi i k x/2^n} |x\rangle \right) = \frac{1}{2^n} \sum_{x=0}^{2^n-1} \sum_{k=0}^{2^n-1} e^{\frac{-2\pi i k}{2^n}(x - 2^n\theta)} |x\rangle. \tag{16}$$

This can be rewritten as follows:

$$QPE(|0\rangle^{\otimes n} \otimes |\psi\rangle) = \frac{1}{\sqrt{2^n}} \left(\left(\left(1 + e^{i\theta_\psi 2^{n-1}}\right)|0\rangle + \left(1 - e^{i\theta_\psi 2^{n-1}}\right)|1\rangle \right) \otimes \left(\left(1 + e^{i\theta_\psi 2^{n-2}}\right)|0\rangle + \left(1 - e^{i\theta_\psi 2^{n-2}}\right)|1\rangle \right) \otimes \right.$$
$$\left. \cdots \otimes \left(\left(1 + e^{i\theta_\psi 2^{0}}\right)|0\rangle + \left(1 - e^{i\theta_\psi 2^{0}}\right)|1\rangle \right) \right) \otimes |\psi\rangle. \tag{17}$$

Appendix B: Hadamard test

Here, we will look at the Hadamard test [15], a method used to create a random variable. Given a unitary matrix U and a state $|\psi\rangle$, the Hadamard test provides an estimate of $Re(\langle\psi|U|\psi\rangle)$. The circuit in **Figure 5** is used for this purpose.

The probability of measuring $|0\rangle$ is

$$P(|0\rangle) = \frac{1}{2}(1 + Re(\langle\psi|U|\psi\rangle)) \tag{18}$$

and the probability of measuring $|1\rangle$ is

$$P(|1\rangle) = \frac{1}{2}(1 - Re(\langle\psi|U|\psi\rangle)). \tag{19}$$

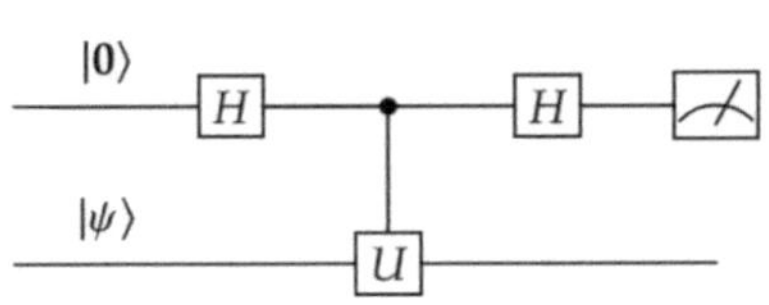

Figure 5.
Circuit implemented for Hadamard test and estimating the probability of |0⟩ output.

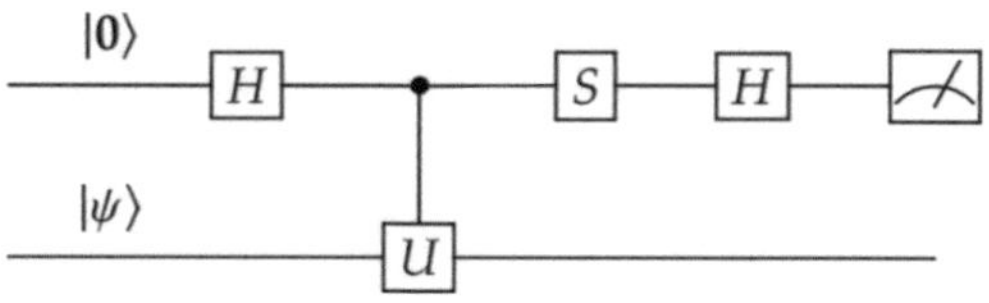

Figure 6.
Circuit implemented for Hadamard test and estimating the probability of $|1\rangle$ *output.*

Thus, $Re\,(\langle\psi|U|\psi\rangle)$ can be estimated by computing $P(|0\rangle) + P(|1\rangle)$, that is the same as $2P(|0\rangle) - 1$. Moreover, $Im(\langle\psi|U|\psi\rangle)$ can be estimated using a similar circuit in **Figure 6**.

This test can also be thought of as a coin flip experiment. Assume M samples with K heads and $M - K$ tails. Then the probability of heads, $P(H) = \frac{K}{M}$. Note that $P(H)$ corresponds to the estimator, $P(|\hat{0}\rangle)$ and thus the estimate of $Re\,(\langle\psi|U|\psi\rangle)$ is $\frac{2K}{M} - 1$.

Author details

Rutuja Kshirsagar
Fujitsu Research of America, Santa Clara, California, United States

*Address all correspondence to: rutujak@vt.edu

References

[1] Deutsch D, Jozsa R. Rapid solution of problems by quantum computation. Proceedings of the Royal Society of London. Series A: Mathematical and Physical Sciences. 1992;**439**:553-558

[2] Shor PW, Algorithms for quantum computation: discrete logarithms and factoring, Proceedings 35th Annual Symposium on Foundations of Computer Science. Santa Fe, NM, USA. 1994. pp. 124-134. DOI: 10.1109/SFCS.1994.365700

[3] Grover LK. A fast quantum mechanical algorithm for database search. In: Proceedings of the twenty-eighth annual ACM symposium on Theory of Computing 1996 (STOC '96). New York, NY, USA: Association for Computing Machinery; 212-219. DOI: 10.1145/237814.237866

[4] Shor PW. Scheme for reducing decoherence in quantum computer memory. Physical Review. A, Atomic, Molecular, and Optical Physics. 1995; **52**(4):R2493-R2496

[5] Steane AM. Active stabilization, quantum computation, and quantum state synthesis. Physical Review Letters. 1996;**78**:2252-2255

[6] Knill E, Laflamme R. Concatenated quantum codes. arXiv preprint quant-ph/9608012. 1996

[7] Peruzzo A et al. A variational eigenvalue solver on a photonic quantum processor. Nature Communications. 2014;**5**:4213. DOI: 10.1038/ncomms5213

[8] Moll N, et al. Quantum optimization using variational algorithms on near-term quantum devices. Quantum Science and Technology. 2018;**3**(3): 030503

[9] Aharonov D, Ben-Or M. Fault-tolerant quantum computation with constant error rate. SIAM Journal on Computing. 1996;**38**:1207-1282

[10] Knill E et al. Resilient Quantum Computation. Science. 1998;**279**: 342-345. DOI: 10.1126/science.279.5349.342

[11] Kitaev AY. Fault tolerant quantum computation by anyons. Annals of Physics. 1997;**303**:2-30

[12] Kitaev AY. Quantum measurements and the abelian stabilizer problem. arXiv preprint quant-ph/9511026. 1995

[13] Brassard G, Høyer P. An exact quantum polynomial-time algorithm for Simon's problem. Proceedings of the Fifth Israeli Symposium on Theory of Computing and Systems. 1997:12-23

[14] Katabarwa A et al. Early fault-tolerant quantum computing. arXiv preprint arXiv:2311.14814. 2023

[15] Aharonov D et al. A polynomial quantum algorithm for approximating the Jones polynomial. Algorithmica. 2005;**55**:395-421

[16] Kimmel S, et al. Robust calibration of a universal single-qubit gate set via robust phase estimation. Physical Review A. 2015;**92**(6):062315

[17] Kshirsagar R, Katabarwa A, Johnson PD. On proving the robustness of algorithms for early fault-tolerant quantum computers. arXiv preprint arXiv:2209.11322. 2022

[18] Liang Q et al. Modeling the performance of early fault-tolerant quantum algorithms. arXiv preprint arXiv:2306.17235. 2023

DOI: http://dx.doi.org/10.5772/intechopen.1004416

[19] Meier AM, et al. Testing the robustness of robust phase estimation. Physical Review A, 2019;**100**(5):052106

[20] Ding Z, et al. Robust Ground-State Energy Estimation under Depolarizing Noise. 2023. Available from: https://api.semanticscholar.org/CorpusID:260091748

Section 2

Quantum Communication and Security

Chapter 4

Quantum Key Distribution Approaches

Abdulbast Abushgra

Abstract

Irrespective of the duration for which the classical system based on bits remains secure, the imminent advent of quantum systems demands a robust cryptographic mechanism to safeguard both manifest data and the existing system's structure and infrastructure against various threats. Quantum mechanics may provide a solution in the form of quantum key distribution (QKD), a set of processes designed to generate and exchange cryptographic keys between two parties. Each unique method and algorithm within QKD can be characterized as a protocol. In 1984, Charles Bennett and Gilles Brassard introduced the first QKD protocol, BB84. Subsequent QKD protocols have been developed based on different principles, including Heisenberg's uncertainty principle, polarization, entanglement, and the non-cloning theory. Other scientists have enhanced some QKD protocols, leading to variations with different names. However, not all these protocols can be implemented in classical systems, as they specifically require a quantum system. The current challenge revolves around the complex and pressing need to develop a QKD protocol that can function effectively on classical systems, quantum systems, or both. In the chapter, QKD protocols are comprehensively examined, with each protocol being meticulously reviewed through technical steps elucidating how the protocol facilitates the exchange of secret keys. Each QKD protocol is dissected to unveil its underlying mechanisms, providing a detailed exploration of the cryptographic procedures involved in generating and exchanging secure cryptographic keys.

Keywords: cryptography, authentication, quantum bits (qubits), plaintext, entanglement, superposition

1. Introduction

Managing the security of assets and servers becomes increasingly challenging as an enormous amount of data is processed every second, with flows tracked down to the millisecond across intricate systems of domains, clusters, subsystems, and platforms. This challenge is anticipated to intensify, reaching a critical security level, particularly when quantum systems become publicly accessible, or possibly even before their widespread use. The classical system, relying on bits and low electricity voltages, can effectively handle a specific capacity of data flow. However, this capacity is constrained by various system requirements, including hardware capabilities.

Moreover, the current classical system used in computing, servers, operating systems, routers, switches, and mobile devices, boasts robust security parameters. These parameters comprise a blend of algorithms and methods that govern the security of the entire system across various layers.

Cryptography has long been the cornerstone of security in classical systems, providing vital protection for sensitive data. However, the challenge for cryptographers has never ceased, especially as the volume and value of shared and exchanged data continue to increase. As data becomes denser and more valuable, the need for robust cryptographic techniques becomes ever more pressing to safeguard against potential threats and vulnerabilities. Cryptographers must continually innovate and evolve their methods to stay ahead of malicious actors and ensure the integrity, confidentiality, and authenticity of data in today's digital landscape. The CIA triad (Confidentiality, Integrity, and Availability) in cybersecurity associated with information systems is guaranteed by the science of cryptography. The classical system of binary is still secure under the protection of number complexities, which difficult algorithms make data exchange more unreachable by hackers or abusers. Time execution to break an RSA (Ron Rivest, Adi Shamir, and Leonard Adleman) [1] encryption key or to conflict an SHA 3 digest [2] probably cost many years of applying brute-force attacks based on the binary system. However, this theory can be changed in terms of using another non-binary system. This non-binary system is an expected quantum system that relies on the law of physics.

Quantum systems, or quantum computers, utilize properties of physics or quantum mechanics that are more powerful in computations and solving problems. This power increases every time and is taken from the nature of photons or the element where a quantum system can be measured. This element is called a quantum bit (qubit), which is the scale of measuring the quantum system parameters. In the quantum dictionary, several triggers enhance the quantum bits measurements, such as polarization, momentum, and mass. All these observable properties can provide a certain way to measure or calculate complex problems. In the field of data security, or more precisely, in cryptography, the most common property that has been used for decades is photon polarization, which the system relies on measuring the state of each photon.

In the realm of quantum systems, the fundamental unit of measurement is the qubit. The stability and efficacy of exchanging secret keys within quantum systems are contingent upon the quality and reliability of the quantum equipment and tools employed. The number of qubits utilized in the system directly influences the computational power of the quantum system; as the quantity of qubits increases, so does the system's capacity for computation. Therefore, maximizing the number of qubits and ensuring the stability of quantum equipment are crucial factors in enhancing the strength and efficiency of secret key exchange processes in quantum systems.

Although the usability of photon polarization requires equipment that must keep and treat the state of photon polarization in a certain case, in this chapter will discuss and focus on the quantum cryptography algorithms regardless of the physical engineering aspects. Specifically, the chapter will conclude the mechanism of quantum key distribution that is considered the tool when a quantum system approaches. The chapter contains an introduction to quantum key distribution (QKD), the properties of QKD, and some flagship protocols that may provide an obvious image of the technique of securing quantum systems.

2. Quantum cryptography

For a long time, digital security has been designed to prevent any negative impact or other circumstances on the system. When this system is exchanging data between parties mostly via the Internet, the challenge becomes dramatically harder because of the natural structure of the Internet. Intruders mostly would intercept any communications to steal or damage the data exchanged. In the classical system that is based on the binary format, there are several techniques used to secure communications between two parties. Let us call these parties Alice (sender) and Bob (receiver). The fundamentals of retaining these exchanges secure are some mathematical complexities that have been used for decades if not centuries. Pseudo Random Number Generation (PRNG) [3] is one of these techniques that has been used for many years, and it is still alive nowadays. The PRNG obtains its strength from the ability to generate a sequence of numbers that are difficult to guess or brute-forced. Brute-force attacks are the only way to break this mechanism, which is still difficult based on the time execution of the used machine (computers). As long as, the developers invest in the speed of computers and their hardware, cryptographers and mathematicians raise the complexity and difficulty to match this equipment.

The actual concern is what if the rational balance between the current system tools and the methods of security interfered with an expected challenge. Three case scenarios can explain these scenarios. Case one is when the system tools and applications are in balance with the security methods used to protect some aspects of the system. This can be fair, but the challenge of keeping the system secure needs extensive surveillance and monitoring. Updating and upgrading the system is an urgent procedure to stay away from any vulnerabilities. Case two can be represented if the security methods exceed the development of the system hardware and software. Theoretically, this case scenario is the most preferred one, specifically users can trust these algorithms. However, the third case scenario is probably the worst because when the system is more powerful than algorithms and security methods, it will be a real challenge to utilize the system, where users are undoubtedly about their assets not secured. In addition, owning and using supercomputers may cause exploitation and data loss. Therefore, quantum computing may cause this shortage and unbalanced situations if cryptographers and scientists cannot reach a powerful mechanism to stand against this power.

Quantum Cryptography theoretically is the ideal solution to protect the classical and quantum systems. Quantum Key Distribution is a mechanism of creating a secret key between two parties. The created secret key relies on quantum mechanics not on mathematical complexity as in the classical system. The quantum mechanics properties guarantee exchanging any qubits between Alice and Bob with an obvious detection of any interception. Because each photon is counted, most of the Quantum Key Distribution protocols adopt the One-Time Pad (OTP) method [4], where the key must be equivalent to the length of the plaintext.

3. Quantum key distribution

The mechanism of distributing a shared key between two or more parties is referred to and is well-known as key distribution. The key distribution is known in the classical cryptographic system as key management too. The key distribution provides a dedicated space where communications can initiate sessions based on the request to share or deliver data such as wireless communications, client and server, and other applications. These sessions reflect several communications between Alice and Bob to create the shared key. The communications may take a few steps, which are based on the protocol design. These steps require a stable connection between the client and server, as well as these steps should not be accessible or simple to break by third parties.

In this section, a compilation of well-known Quantum Key Distribution (QKD) protocols [5] is provided, each representing a distinct method or technique for generating a secret key. All QKD protocols involve quantum communications conducted through a quantum channel, and some may additionally require a classical channel. The capacity of the quantum channel is still not clearly defined, as most experiments have been conducted either in classical settings or within the confines of constrained quantum systems. This lack of an explicit capacity arises from the intricate nature of quantum communication and the challenges associated with accurately quantifying the capabilities of quantum channels. Ongoing research endeavors to explore and better understand the potential capacities of quantum channels. The outlined QKD protocols focus on illustrating the methods for establishing a secret key between two parties, without an explicit intent to demonstrate the physical capacities of quantum channels. It's important to acknowledge that quantum channels inherently exhibit noise, making them susceptible to errors during data transmission.

The quantum channel has versatile applications; it can facilitate classical information transmission by utilizing pure orthogonal states. Additionally, the quantum channel can be employed to transmit nonorthogonal states and even quantum entanglement. The pure orthogonal states are quantum states that are completely distinct from each other and have no overlap in their properties, whereas nonorthogonal states share some common characteristics and exhibit overlap in their properties [6]. The approach to information transmission also hinges on whether the input comprises unentangled or entangled quantum states. Consequently, in the realm of quantum channels, numerous new capacity definitions emerge when compared to classical communication channels. The richness of quantum states and their entanglement properties introduces a spectrum of possibilities and complexities that distinguish quantum channels from their classical counterparts.

3.1 BB84 protocol

The most common QKD protocol was published in 1984 [7], by two physicists Charles Bennett and Gilles Brassard. After a while, the quantum protocol became a flagship in quantum cryptography, which is called BB84 referred to by the last names of the authors. The theory of BB84 concluded in a simple format that represents digital information as polarized photons, where each classical bit (0, 1) matches a specific position of the photon state ($|0\rangle, |1\rangle$). In this protocol, the information (plaintext) is encoded based on two bases (rectilinear and diagonal) and four states (polarization directions as, 0, 45, 90, and 135 degrees). Other physical equipment may be needed as well as extra processes that could affect the precise measurements. However, in this chapter, the main concept relies on explaining the algorithms of the QKD protocol and its mechanism.

The operational principle of the BB84 protocol involves emitting a sequence of polarized light through a polarizing apparatus. In cryptographic terms, the transmission of polarized light serves as a source of random bits initiated over a quantum channel. These random bits are generated by quantum bits (Qubits), with each qubit having a defined position in space. The measurement of these photons is governed by the principles of superposition theory. The BB84 protocol leverages the properties of quantum superposition to establish a secure communication channel by encoding information in the polarization states of individual qubits.

Generating a secret key through the BB84 protocol necessitates both communicating parties, *i.e.*, the sender and receiver, to possess a random number generator strategically positioned between them. This generator can be centrally located between legitimate parties. In the initial stages, the sender (Alice) begins by creating plaintext X, which is then transformed into a bit string. Concurrently, Alice initializes a random set of bases (horizontal or diagonal) that corresponds in length to the plaintext X. These bases encompass four states ($|+\rangle$, $|-\rangle$, $|0\rangle$, and $|1\rangle$), with each state in a different basis representing the probability of 0 or 1. Moreover, the entire set of prepared states $|\varphi_i\rangle$ is transmitted through a quantum channel, maintaining the same polarization as the prepared state, provided there is no interference.

The security of the submitted qubits in the BB84 protocol hinges on the non-cloning theorem and the Heisenberg uncertainty principle. The non-cloning theorem is a consequence of the superposition principles within quantum mechanics. Additionally, the non-cloning property enhances the stability of the BB84 protocol by identifying potential attacks, even in the face of persistent attempts by attackers to breach cryptographic protocols. The Heisenberg uncertainty principle, on the other hand, posits the impossibility of simultaneously preparing or measuring states in a given environment under quantum conditions, particularly with respect to position and momentum.

In a broader context, quantum key distribution protocols can be classified based on two aspects of photon behavior: one relies on superposition states (orthogonal/non-orthogonal), and the other is based on entangled states, with the BB84 protocol utilizing polarized orthogonal states. In the case of superposition states, Alice transmits a state generated on the basis of either $(\times)$ or $(+)$, as mentioned earlier. In this scenario, Bob randomly chooses to work on one of these bases. For instance, if Alice utilizes the $(\times)$ basis to send a $|1\rangle$ state, she will transmit a $|\nwarrow\rangle$ state. Similarly, if she intends to send a $|\uparrow\rangle$ state, and Bob has already measured the $|\uparrow\rangle$ state in the $(+)$ basis, he will record a $|1\rangle$ state. Also, if Alice sends a photon as $|\nearrow\rangle$ or $|\nwarrow\rangle$ state and Bob just measures the photon in the basis $(+)$, the measurement will be in the polarized states in Eq. (1) as follows [5]:

$$\begin{aligned}
|\nwarrow\rangle\ with\ (+) &= \frac{1}{\sqrt{2}}(|\uparrow\rangle - |\rightarrow\rangle),\\
|\nearrow\rangle\ with\ (+) &= \frac{1}{\sqrt{2}}(|\uparrow\rangle + |\rightarrow\rangle),\\
|\uparrow\rangle\ with\ (\times) &= \frac{1}{\sqrt{2}}(|\nearrow\rangle + |\nwarrow\rangle),\\
|\rightarrow\rangle\ with\ (\times) &= \frac{1}{\sqrt{2}}(|\nearrow\rangle - |\nwarrow\rangle).
\end{aligned} \tag{1}$$

The potential outcomes of measuring polarization states in the BB84 protocol can be represented in the Bloch sphere. The Bloch sphere is a geometrical representation of the state space of a two-level quantum system, such as a qubit, in three-dimensional space (x, y, and z axes) [8]. In the context of the BB84 protocol, each qubit transmitted by Alice can be in one of four possible states: horizontal polarization $|H\rangle$, vertical polarization $|V\rangle$, diagonal polarization $|D\rangle$, or anti-diagonal polarization $|A\rangle$. These states can be visualized as points on the Bloch sphere, where different measurements of polarization correspond to different rotations of the Bloch vector representing the qubit's state.

Indeed, the establishment of a Shared Secret Key through the BB84 protocol involves several sequential steps for both parties. These steps are outlined as follows:

Step 1: Alice initiates the process by configuring the length of plaintext X to form a string of n-bits. Subsequently, these n-bits are then applied to a randomly prepared basis, which could be either $(\times)$ or $(+)$ (**Table 1**).

Step 2: For each randomly chosen basis, a corresponding random state is generated. If the basis is $|\times\rangle$, the outcome will be either $|0\rangle$ or $|1\rangle$. Likewise, if the basis is $|+\rangle$, the result will be either $|0\rangle$ or $|1\rangle$, as shown in **Table 2**.

Step 3: Upon Alice submitting the string of n-qubits, Bob proceeds to measure these incoming n-qubits using randomly selected bases. Subsequently, Bob acquires a string of states that corresponds to n-bits. In cases where Bob is unable to measure all the submitted qubits successfully, both parties collaboratively release additional qubits by sharing the used bases through a public channel.

Step 4: Both Alice and Bob engage in the assessment of potential errors introduced by Eve. In the BB84 protocol, various error correction methods are employed for this purpose. The raw secret key undergoes processing as Alice and Bob compare matching bits, with uncorrelated bits being discarded. This process, known as the sifting procedure, serves to fortify the security against Eve's attempts to gather information and detect any errors (**Table 3**).

Binary numbers	0	1	2	3	4	5
Random Bits	0	1	1	0	1	1
Random Bases	+	+	×	+	+	+
Submitted Qubits	→	↑	↖	→	↑	↑

Table 1.
Alice transmits a string of n random bits filtered through random bases [5].

Binary numbers	0	1	2	3	4	5
Random Bases	×	+	×	×	+	×
Observed Qubits	↗	↑	↖	↖	↑	↗
Measured Bits	0	1	1	1	1	0

Table 2.
Bob receives a string of n random bits measured in random states [5].

Binary numbers	0	1	2	3	4	5
Random bases for A	+	+	×	+	+	+
Random bases for B	×	+	×	×	+	×
Agreement		✓	✓		✓	
Exchanged Key		1	1		1	

Table 3.
Alice and bob compare generated bases [5].

Step 5: Following the comparison of sent and received qubits, the communication progresses to the reconciliation phase only when the error rate is low. Conversely, if the error rate is deemed excessively high, Alice and Bob terminate the ongoing communication.

Step 6: In the event of a low error rate, Alice and Bob proceed to share the raw key. The raw key comprises the matched qubits from both parties, and any unmatched qubits are slated for removal from the shared key.

Step 7: Subsequently, Alice and Bob initiate the correction of erroneous qubits in a distinct phase, illustrated in **Table 4**, as they work towards minimizing the number of vulnerable qubits.

Step 8: Upon error checking, Alice and Bob exchange a Shared Secret Key, matching the length of plaintext X. Notably, in this phase, Alice could potentially attempt deception by sending a different basis (rectilinear or diagonal, or neither) such that she cannot align with any of Bob's table records from step (3). In contrast, Bob's table records reflect the outcomes of probabilistic behavior that are beyond the control of the matched raw key.

It is crucial to recognize that if Alice attempts to deceive in step (1), such as by transmitting a combination of rectilinear and diagonal states, she forfeits the capability to align with Bob's table records following step (1). In conclusion, the BB84 protocol is deemed a secure protocol, and it stands out for its simplicity relative to contemporary Quantum Key Distribution (QKD) protocols. This simplicity is rooted in the fundamental laws of physics governing key generation [5]. Indeed, while the BB84 protocol may not be the only secure Quantum Key Distribution (QKD) protocol, and its security may be subject to certain limitations, its significance in the field of quantum cryptography cannot be overstated. Much like the Data Encryption Standard (DES) in classical cryptography and the Open Systems Interconnection (OSI) model in networking infrastructure, the BB84 protocol has become a foundational element in the science of quantum communication.

The BB84 protocol offers a straightforward and intuitive framework for secure key exchange between two parties within a quantum system. Its simplicity and clarity make it a valuable starting point for understanding the principles of quantum cryptography, providing a basic blueprint for implementing secure communication protocols in quantum environments. While other QKD protocols may offer enhanced security features or better performance under certain conditions, the BB84 protocol remains a fundamental reference point in the study and development of quantum communication technologies. Its inclusion in the arsenal of quantum cryptographic techniques underscores its importance as a cornerstone in the quest for secure communication in quantum systems.

Binary numbers	0	1	2	3	4	5
Exchanged key (R1)		1	1		1	
Randomly chosen			✓		✓	
Exchanged key (R2)		1	1		1	
Final Agreement			✓		✓	
Unrevealed secret key		1				

Table 4.
Alice and bob evaluate the remaining bits in public [5].

3.2 AK15 protocol

In 2017, Abdulbast Abushgra and Khaled Elleithy published [9] a quantum key distribution protocol that is based on matrix mechanism and entanglement states. This protocol introduces an enhanced Quantum Key Distribution (QKD) scheme, incorporating user authentication within an entangled channel. The featured QKD algorithm is technically structured into two quantum channels: the first channel being an EPR channel (entangled states channel), and the second channel being a quantum channel (qubit channel in superposition) [10]. The proposed QKD scheme is designed to conclude if the necessary authentication between the communicating parties of the EPR channel is unsuccessful. The AK15 protocol contains several phases such as preparation phase, submission phase, mathematical phase, and measurement phase.

The EPR channel, named after Einstein, Podolsky, and Rosen, who initially explored it in a renowned paper in 1935 [11], denotes a theoretical framework within quantum mechanics centered around entangled states. Entanglement emerges when the properties of particles become correlated to such an extent that the state of one particle is reliant on the state of another, irrespective of the spatial separation between them. The EPR channel holds substantial importance in comprehending quantum phenomena like quantum teleportation and quantum cryptography, where entangled states serve as a mechanism for information transfer and ensuring secure communication.

This protocol encompasses two vital methods, exclusively undertaken by Alice, who functions as one of the communicators or the sender. Additionally, these methods complement each other reciprocally, with the understanding that the first preparation is never authorized without establishing the second preparation and vice versa. One crucial assumption in this proposed Quantum Key Distribution (QKD) scheme is that Alice, as the trusted sender party, provides accurate information without engaging in deceptive practices. To establish a Shared Secret Key (SSK), Alice, serving as one of the communicators or the sender, must possess knowledge of an original plaintext X intended for transfer to Bob, who acts as the receiver. In the initial steps, the original plaintext X is transformed into a string of bits (data). Subsequently, Alice converts this string of bits into quantum bits (qubits). The conversion involves passing each individual bit (bit $\in \{0, 1\}$) through a qubit converter, transitioning from a regular bit to a unit vector in the Hilbert space ℓ^2 [9].

$$\text{Plaintext X} \approx \text{bits} \in \{0,1\}^n \tag{2}$$

$$\{0.1\}^n \in \begin{cases} |0\rangle, |\varphi\rangle = (\alpha|0\rangle + \beta|1\rangle) \\ \text{OR} \\ |1\rangle, |\varphi\rangle = (\alpha|0\rangle - \beta|1\rangle) \end{cases} \tag{3}$$

$$DM_A = \log_2 n \tag{4}$$

In scenarios where the entire data transmission process involves quantum bits (qubits) without the need for conversion from classical binary bits, certain steps of the AK15 protocol can be streamlined or omitted. Specifically, the conversion phase from classical to quantum bits can be eliminated, leading to potential time improvements in the protocol's execution. Moreover, the absence of the conversion phase can present advantages for utilizing the AK15 protocol across various systems, including classical, post-quantum, and full quantum systems. By operating solely with qubits, the protocol becomes agnostic to the underlying system architecture, allowing for seamless

$\|\emptyset_{AB}\rangle$					
$\|\emptyset_{AB}\rangle$	$\|\emptyset_{AB}\rangle$				
$\|\emptyset_{AB}\rangle$	$\|\emptyset_{AB}\rangle$	$\|\emptyset_{AB}\rangle$			
$\|\emptyset_{AB}\rangle$	$\|\emptyset_{AB}\rangle$	$\|\emptyset_{AB}\rangle$	$\|\emptyset_{AB}\rangle$		
$\|\emptyset_{AB}\rangle$	$\|\emptyset_{AB}\rangle$	$\|\emptyset_{AB}\rangle$	$\|\emptyset_{AB}\rangle$	$\|\emptyset_{AB}\rangle$	

Table 5.
The prepared matrix DM after calculating the length of X into three sections is: ($|\emptyset_{AB}\rangle$ lower triangle, $|\varphi_{AB}\rangle$ upper triangle, and $|\omega_{AB}\rangle$ diagonal line, where the upper-triangle equals the lower-triangle (notice: There is no similarity in the DM cells, it is just to show the differentiation between the matrix sections) [9].

integration into different environments without the need for additional adaptations or conversions. Overall, eliminating the conversion phase enhances the efficiency and versatility of the AK15 protocol, making it well-suited for a wide range of applications in both classical and quantum communication systems (**Table 5**).

In this context, DM represents the dimensions of the prepared matrix, typically with an equal number of columns and rows. The parameter 'n' corresponds to the length of the converted plaintext in terms of n-bits that Alice intends to share with Bob.

Additionally, Alice proceeds to populate the upper triangle of the matrix (excluding the diagonal line) with random qubits, represented as well-known bits converted to qubit states (e.g., $|\varphi\rangle$), as illustrated in **Table 6**. These randomly generated qubits serve as decoy states during the reconciliation phase between the legitimate users (**Table 7**).

It's considered that in the proposed protocol, the reconciliation phase is integrated into the initial phase of communication, rather than being a separate phase as observed in several Quantum Key Distribution (QKD) protocols. The matrix is fully prepared, with the exception of the diagonal line. Alice adjusts the cells along the diagonal line based on the summation of each row, as depicted in Eq. (5). If the sum of a chosen row is odd, Alice adds the state $|1\rangle$ to the empty cell ($\times$) to ensure the row becomes even (e.g., Eq. (3)). Conversely, if the row sum is even, Alice adds the state $|0\rangle$ to the matrix cell. Consequently, Alice prepares the entire matrix with even rows, as shown in **Table 8**.

The diagonal line serves as confirmation between the legitimate users, acting as parity cells and offering additional protection against Photon-Number Splitting (PNS)

	$\vert\varphi_{AB}\rangle$	$\vert\varphi_{AB}\rangle$	$\vert\varphi_{AB}\rangle$	$\vert\varphi_{AB}\rangle$	$\vert\varphi_{AB}\rangle$
		$\vert\varphi_{AB}\rangle$	$\vert\varphi_{AB}\rangle$	$\vert\varphi_{AB}\rangle$	$\vert\varphi_{AB}\rangle$
			$\vert\varphi_{AB}\rangle$	$\vert\varphi_{AB}\rangle$	$\vert\varphi_{AB}\rangle$
				$\vert\varphi_{AB}\rangle$	$\vert\varphi_{AB}\rangle$
					$\vert\varphi_{AB}\rangle$

Table 6.
The prepared matrix DM after calculating the length of X into three sections: $|\varnothing_{AB}\rangle$ lower triangle, $|\varphi_{AB}\rangle$ upper triangle, and $|\omega_{AB}\rangle$ diagonal line, where the upper-triangle equals the lower-triangle (notice: There is no similarity in the DM cells, it is just to show the differentiation between the matrix sections) [9].

$\vert\omega_{AB}\rangle$	$\vert\varphi_{AB}\rangle$	$\vert\varphi_{AB}\rangle$	$\vert\varphi_{AB}\rangle$	$\vert\varphi_{AB}\rangle$	$\vert\varphi_{AB}\rangle$
$\vert\emptyset_{AB}\rangle$	$\vert\omega_{AB}\rangle$	$\vert\varphi_{AB}\rangle$	$\vert\varphi_{AB}\rangle$	$\vert\varphi_{AB}\rangle$	$\vert\varphi_{AB}\rangle$
$\vert\emptyset_{AB}\rangle$	$\vert\emptyset_{AB}\rangle$	$\vert\omega_{AB}\rangle$	$\vert\varphi_{AB}\rangle$	$\vert\varphi_{AB}\rangle$	$\vert\varphi_{AB}\rangle$
$\vert\emptyset_{AB}\rangle$	$\vert\emptyset_{AB}\rangle$	$\vert\emptyset_{AB}\rangle$	$\vert\omega_{AB}\rangle$	$\vert\varphi_{AB}\rangle$	$\vert\varphi_{AB}\rangle$
$\vert\emptyset_{AB}\rangle$	$\vert\emptyset_{AB}\rangle$	$\vert\emptyset_{AB}\rangle$	$\vert\emptyset_{AB}\rangle$	$\vert\omega_{AB}\rangle$	$\vert\varphi_{AB}\rangle$
$\vert\emptyset_{AB}\rangle$	$\vert\emptyset_{AB}\rangle$	$\vert\emptyset_{AB}\rangle$	$\vert\emptyset_{AB}\rangle$	$\vert\emptyset_{AB}\rangle$	$\vert\omega_{AB}\rangle$

Table 7.
The entire prepared matrix DM_A at Alice's side is ready to be converted to one string I_{QUBIT} [9].

$\lvert\omega_{AB}\rangle$	$\lvert\varphi_{AB}\rangle$	$\lvert\varphi_{AB}\rangle$	$\lvert\varphi_{AB}\rangle$	$\lvert\emptyset_{AB}\rangle$	$\lvert\varphi_{AB}\rangle$
$\lvert\emptyset_{AB}\rangle$	$\lvert\varphi_{AB}\rangle$	$\lvert\varphi_{AB}\rangle$	$\lvert\varphi_{AB}\rangle$	$\lvert\omega_{AB}\rangle$	$\lvert\varphi_{AB}\rangle$
$\lvert\emptyset_{AB}\rangle$	$\lvert\emptyset_{AB}\rangle$	$\lvert\omega_{AB}\rangle$	$\lvert\varphi_{AB}\rangle$	$\lvert\varphi_{AB}\rangle$	$\lvert\varphi_{AB}\rangle$
$\lvert\emptyset_{AB}\rangle$	$\lvert\omega_{AB}\rangle$	$\lvert\emptyset_{AB}\rangle$	$\lvert\varphi_{AB}\rangle$	$\lvert\varphi_{AB}\rangle$	$\lvert\varphi_{AB}\rangle$
$\lvert\emptyset_{AB}\rangle$	$\lvert\emptyset_{AB}\rangle$	$\lvert\varphi_{AB}\rangle$	$\lvert\emptyset_{AB}\rangle$	$\lvert\omega_{AB}\rangle$	$\lvert\varphi_{AB}\rangle$
$\lvert\emptyset_{AB}\rangle$	$\lvert\omega_{AB}\rangle$	$\lvert\emptyset_{AB}\rangle$	$\lvert\varphi_{AB}\rangle$	$\lvert\emptyset_{AB}\rangle$	$\lvert\emptyset_{AB}\rangle$

Table 8.
The received qubits were inserted from left to right and up to down sequentially (DM_B) by the receiver. The qubits will be mismatched. With the prepared matrix (DM_A), which is impossible to be detected by an eavesdropper [9].

attacks. However, when Bob measures the incoming qubits, he gains insights into whether these qubits experienced interruptions from either eavesdroppers or the environment.

$$I_{ROW3} \in \{|1\rangle,\ 0\rangle.\ \ |\times\rangle.\ \ |0\rangle.\ \ |1\rangle.\ \ |1\rangle\ |\} \tag{5}$$

Each row in the prepared matrix, denoted as I_{ROW} (as illustrated in, for example, Eq. (3)), is configured to have an even total summation of elements. Consequently, the empty cell (diagonal cell, e.g., $(\times)$) is assigned the state $|0\rangle$ *or* $|1\rangle$. Applying Eq. (3), the diagonal state $|\times\rangle$ is designated as the state $|0\rangle$ because the total of $|1\rangle$ states is now equal to the total of $|0\rangle$ states. This procedure is consistently applied to the entire designed matrix DM (DMA $\in \{I_{ROW1}, I_{ROW2}, I_{ROW3}, \ldots, I_{ROWN}\}$) [9].

As a result, Alice possesses a set of rows (indexed as I_{ROWN}) that will later be transmitted to Bob during a quantum channel. This submission involves randomly selecting indices of matrix rows from the prepared matrix DM and sequentially inserting each row, based on the randomly chosen indices, into a single string referred to as I_{QUBIT}. Subsequently, the string of qubits I_{QUBIT} is transmitted into a quantum channel, where each prepared qubit is polarized into a superposition state ($|\times\rangle$ *or* $|+\rangle$).

Ultimately, the earlier string of qubits, I_{QUBIT}, cannot be effectively utilized until the EPR communication is prepared through the entangled channel (EPR channel). Given that Alice has already obtained sufficient details from the prepared data in the DMA, she proceeds to initiate another phase based on qubit preparation. Furthermore, Alice prepares another communication with Bob to validate authentication and facilitate the sharing of reconciliation details. This additional phase enhances the security and reliability of the quantum communication process between Alice and Bob.

3.3 Coherent one-way protocol

The Coherent One-Way (COW) protocol [5, 12] operates through a straightforward mechanism, relying on the decoding of information into specific time slots. Alice transmits coherent pulses in logical states or employs decoy states. Each logical bit is represented as either (μ–0) for logical 0 or (0–μ) for logical 1 through a sequence of two pulses. Additionally, for enhanced security, Alice introduces decoy sequences of (μ–μ) alongside the other logical states. If the pulses sent to the interferometer are precisely aligned on Bob's end, the received pulses will be accurately detected on DM1 (interferometer), with no detection occurring on DM2 (detector). Consequently, any attempt by an eavesdropper to intercept results in a loss of coherence, noticeable on the detector.

$$\begin{aligned} \text{logic 1} &: |0\rangle + |\mu\rangle \\ \text{logic 0} &: |\mu\rangle + |0\rangle \\ \text{Decoy} &: |\mu\rangle + |\mu\rangle, \end{aligned} \tag{6}$$

where μ is the mean photon number per pulse.

In this protocol, the transfer and receipt of data hinge on the timing of signal arrival rather than the polarization of optical signals. The COW protocol operates as follows:

Step 1: Alice transmits a sequence of binary bits to Bob using time slots, generating logical states of $|1\rangle$ or $|0\rangle$, each with an equal probability, unless decoy states are introduced. The probability for each of the $|1\rangle$ or $|0\rangle$ states is ½, and the calculation for adding decoy states involves $\frac{(1-f)}{2}$, where f represents the probability of decoy state generation.

Step 2: Bob utilizes time detection to generate a raw key, employing various detectors for all preceding processes to enhance the security rate in Eq. (7).

$$V = \frac{p(D_{M1}) - p(D_{M2})}{p(D_{M1}) + p(D_{M2})} \tag{7}$$

where $p(D_{Mj})$ represents the probability of clicks at the time corresponding to D_{M1}, as illustrated in **Figure 1**.

Step 3: Bob determines the number of bits through simultaneous procedures involving both the data detector and time detection on his end.

Step 4: While monitoring the detectors, Alice verifies the presence of the sequence of decoy states and bit sequences. If either is absent, it indicates potential

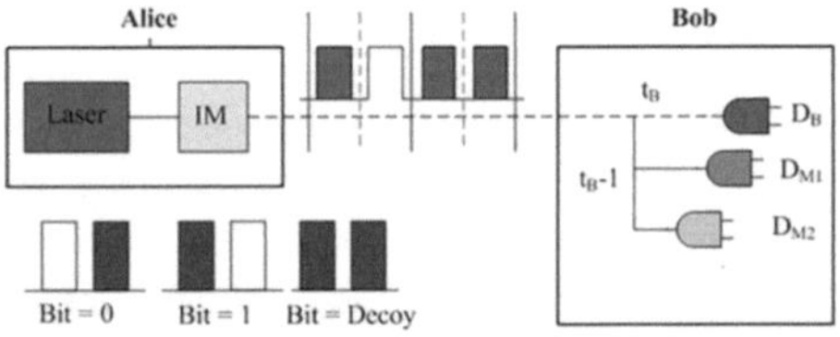

Figure 1.
The coherent-one-way (COW) protocol is established between two trusted entities, Alice and bob, necessitating specific equipment for processing submissions and measuring split times. The submitted qubits comprise three distinct categories of data: The value of 0, the value of 1, and the value of decoy states.

eavesdropping by Eve. In such a scenario, Alice disrupts the coherence by breaking it into two pulses to identify any interruptions in the state.

Step 5: Alice communicates to Bob the exclusion of certain bits from the raw key, specifying that these bits pertain to the decoy state sequence.

Step 6: The secret key is derived by eliminating the decoy sequences from the raw key through a classical process. Subsequently, the shared key is acquired through error correction and privacy amplification.

This protocol, as outlined in the report, is crafted to be a resilient quantum protocol capable of withstanding reduced interference visibility and PNS attacks. The Coherent One-Way protocol boasts straightforward transmissions into data lines, minimal losses at the measurement side, and a low Quantum Bit Error Rate (QBER) detection.

3.4 S13 protocol

S13, introduced by Serna in 2013 [13], is a quantum key distribution protocol. While it aligns with the principles of the BB84 protocol in quantum procedures, it distinguishes itself through variations in the classical channel. S13 was specifically crafted to be seamlessly integrated into existing system devices, eliminating the necessity for any modifications. S13 protocol was designed by a computer scientist, where it was following another QKD protocol that was designed in 2009, known as S09 [14]. The security of S09 was based on a secret key that should be multi-exchanged qubits every single time of connection between Alica and Bob.

Therefore, S13 was designed to eliminate the time consumption and redundancy of the process taken in S09. S13 shares the same quantum communication phase as BB84, a detail that has already been addressed in this section and will not be reiterated. The subsequent phase of the S13 protocol is elucidated as follows [5]:

Phase 1: Quantum part

- Raw key exchange: (as shown in the BB84 protocol).
- Random seed: one of the communicating parties creates a random binary string $(x_1\, x_2 \ldots x_N)$.

Missing key exchange:

1. Alice makes a summation of the random binary string with the binary basis from the first part and obtains a binary basis $(t_1\, t_2 \ldots t_N)$. Alice then randomly generates another string of binary $(j_1 j_2 \cdots j_N)$, where this is an exchanged key with Bob.
2. Bob sums each of the sequences sent to him by Alice with the created state vertical $(1 \bigoplus m_k) \bigoplus x_k$, where $(k = 1, 2 \ldots N)$. Thus, the sum becomes a binary string basis $(n_1\, n_2 \ldots n_N)$. Next, Bob measures the received state$|\Psi_{t_k j_k}\rangle$, with the correspondence of the basis (B_{nk}) to generate $(b_1\, b_2 \ldots b_N)$.

Phase 2: Classical part.

Alice and Bob apply function (f) to different binary exchanges in a set of binary strings:

$$f(z,x,y) := \begin{cases} x, z = 0 \\ y, z = 1 \end{cases} \tag{8}$$

Asymmetric cryptography:

Step 1: Alice sums the binary string created by her in quantum part i with a random string of binary values that were created by missing the key exchange j.

$$i_k \oplus j_k, \ (k = 1, 2 \ldots N), \tag{9}$$

where $(y_1 \, y_2 \ldots y_N)$ will be sent to Bob.

Step 2: To obtain the public key, Bob encrypts:

$$\begin{aligned} u_k &= n_k \oplus f\left(m_k, a_k, b_k, y_k\right), \\ v_k &= n_k \oplus f\left(m_k, b_k, a_k \oplus y_k\right). \end{aligned} \tag{10}$$

Step 3: Alice makes a summation to obtain the private string of m_k, which is:

$$t_k \oplus f\left(s_k, (1 \oplus i_k) \oplus u_k, j_k \oplus v_k\right), \tag{11}$$

and then decrypts the string $(m_1 \, m_2 \ldots m_N)$.

Private Reconciliation:

Step 4: Bob receives the binary sequence $(l_1 \, l_2 \ldots l_N)$ after completing the comparison between $(s_1 \, s_2 \ldots s_N)$ and $(m_1 \, m_2 \ldots m_N)$ by Alice.

Step 5: Bob sums the sequence of bases m_k with l_k, where $(m_k \oplus l_k), k = 1, 2 \ldots N$.

$$\begin{aligned} & f\left(l_k, a_k, b_k \oplus y_k\right) \equiv i_k \\ & f\left(l_k, a_k \oplus y_k, b_k\right) \equiv j_k, \\ & (k = 1, 2 \ldots N). \end{aligned} \tag{12}$$

This is to obtain the private string s_k.

Bob then gets the private string from Alice $(i_1 \, i_2 \ldots i_N)$.

Ultimately, the S13 protocol is tailored to seamlessly integrate with current devices, particularly during the exchange phase following qubit transmission. Numerous exchanges over the public channel could result in time wastage and create opportunities for eavesdropping on data. Moreover, it is noteworthy that S13 represents a significant advancement compared to the S09 protocol [14], which was acknowledged for its complexity as a quantum key distribution protocol.

3.5 KMB09 protocol

The protocol [15], introduced in 2009 by Khan, Murphy, and Beige, is crafted to withstand PNS (Phase-Number-Splitting) attacks. In their description, Khan et al. outline a communication protocol involving two parties, namely Alice and Bob, and a potential eavesdropper named Eve. For the protocol to function securely, both parties must utilize two distinct bases, denoted as e and f. It is crucial that whenever the same basis is employed by both parties, they use different indices, represented by the variable i. Furthermore, the i index is openly disclosed between the legitimate parties.

This index is a shared parameter that can be attributed to Alice's prepared indices as i and Bob's measured indices as j, enhancing the transparency and security of the communication process.

In their work KMB09, the authors aimed to develop a protocol with resilience against Phase-Number-Splitting (PNS) attacks. The motivation behind KMB09's creation was the inadequacy of existing protocols when implemented over relatively short distances, where the system error rate could surpass the detectability threshold of an eavesdropper. The protocol underwent optimization by introducing the use of Index Transmission Error Rate (ITER) instead of Quantum Bit Error Rate (QBER) during the reconciliation phase. The subsequent steps provide a concise overview of the KMB09 protocol as follows:

Step 1: Alice generates a random sequence of classical bits and selects a random index i from the set {1, 2, …, N}.

Step 2: Alice encodes the prepared bits onto single photons in either the $|\,e_i\rangle$ or $|f_i\rangle$ basis and sends them to Bob.

Step 3: Bob randomly measures each incoming state using bases e and f.

Step 4: Alice publicly communicates the randomly chosen indices i.

Step 5: Bob interprets the measurement outcomes based on the received indices.

Step 6: Bob publicly communicates with Alice, confirming the successful reception of photon measurements and the generation of the secret key.

Step 7: Alice and Bob assess the possibility of eavesdropping by Eve using Eq. (5).

$$P_{ITER} = 1 - \frac{1}{2N}\sum_{i=1}^{N}\sum_{k=1}^{N}\left[|\langle g_k|e_i\rangle^4 + \langle g_k|f_i\rangle^4|\right] \tag{13}$$

In the given context, the bases e, f, and g are utilized, where the state $|g_k\rangle$ represents Eve's potential measurement outcomes. This state is transmitted to Bob without any modification.

The polarization of a single photon is initiated in multi-dimensional states, as illustrated in **Figure 2**. These states are established on either orthogonal or non-orthogonal bases.

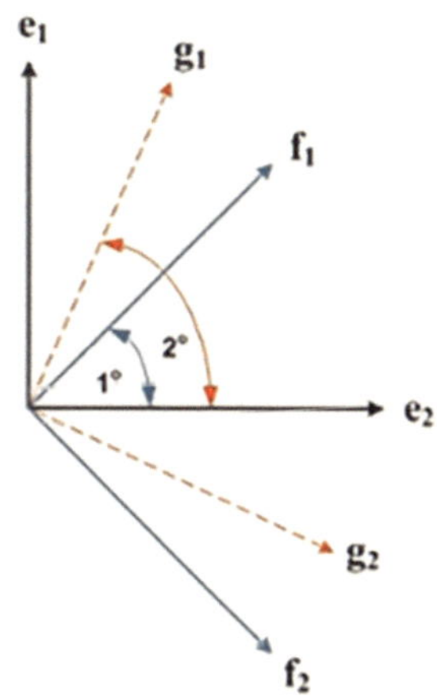

Figure 2.
Only two bases vector used by legitimate users (Alice and bob), and third party (eve), where is the N = 2 protocol.

The KMB09 protocol is specifically formulated to operate under ideal conditions, ensuring that it is infeasible for Alice and Bob to have different indices while employing the same basis. This design enhances the protocol's robustness against potential eavesdroppers attempting to conceal their presence. Moreover, the protocol leverages the strong correlation between Quantum Bit Error Rate (QBER) and Index Transmission Error Rate (ITER), causing the eavesdropper to generate a unique signature that is easily detectable. This attribute contributes to the protocol's effectiveness in detecting and thwarting eavesdropping attempts [5].

4. The future of QKD

Despite the myriad of techniques, methods, and algorithms in the realm of key distribution, the core principles revolve around ensuring the security of data exchange between legitimate parties, as well as the strength and efficiency of the generated keys. The quantum key distribution (QKD) protocols outlined above exemplify a subset of protocols that can be implemented using various techniques and algorithms, all of which leverage photon polarization in the quantum channel. While the BB84 protocol highlights the simplicity of creating and exchanging a secret key between two legitimate parties and remains the most renowned, it is not the sole protocol applicable to classical and quantum systems. Other protocols offer unique approaches to sharing and extracting a string of qubits under diverse conditions, each presenting distinct advantages in the field of quantum cryptography.

As long as the quantum system has not achieved full availability, theoretical applications such as Quantum Key Distribution (QKD) protocols are subject to continuous scrutiny for improvements or revisions. Some of these protocols have been tested on classical or hybrid systems. Additionally, the development of quantum platforms is still ongoing. The primary rationale behind designing these protocols and algorithms before the realization of a full quantum system is to safeguard existing systems against potential attacks leveraging quantum fundamentals. This proactive approach aims to preemptively address security vulnerabilities and ensure the resilience of current systems in anticipation of future advancements in quantum technology.

As a result, this chapter serves as a valuable resource for individuals interested in quantum cryptography, particularly quantum key distribution (QKD). It covers several QKD protocols along with their implementations and theoretical aspects. While the BB84 protocol is widely regarded as a benchmark in this field and is included in the chapter, other protocols such as AK15, COW, S13, and KMB09 are also featured. These protocols were selected because they represent different formats and techniques compared to BB84, providing a comprehensive overview of the diversity within QKD protocols. There are definitely numerous other QKD protocols in existence, and delving into each of them may require more space than a single chapter allows. This suggests a potential avenue for future work, perhaps in the form of a book, to provide an in-depth exploration of the multitude of QKD protocols and their applications.

5. Conclusion

Without any doubt, the value of assets in digital systems is rising every minute, and securing these assets becomes more expensive. The classical system would handle a limited challenge, but it might be a crisis if the quantum computer arrives in public. Utilizing the power of quantum computing without preparation is considered as destroying the manifest digital infrastructure. Quantum Key Distribution is a mechanism for creating a secret key between entities. These entities can share a key with confidence that no interruption has occurred during the key exchange. Although quantum key distribution protocols utilize the law of physics and require unique equipment, several experiments have approved that quantum key distribution can be used over the classical system. This enhances the capability of the classical current system to defend and stand against several threats, such as quantum computing acceleration to solve problems. In this chapter, some quantum key distribution protocols are chosen to explore several algorithms in this field. Also, most of these quantum key distribution protocols are not a one-way function, which means each protocol has a unique design and algorithm.

Acknowledgements

Special thanks to my colleague Leonore Fleming for the unwavering support that greatly contributed to the completion of this work and many others. Additionally, I would like to express my gratitude to my wife Manal for her continuous support throughout this endeavor. I must also extend my appreciation to Lamar, Laryn, Ruseel, Salsabeel, Mohammed, and Saja for their efforts in creating an amazing work environment. Your support and dedication have been instrumental in the success of this project.

Conflict of interest

The author declares no conflict of interest.

Author details

Abdulbast Abushgra
Cybersecurity, Utica University, United States of America

*Address all correspondence to: aaabushg@utica.edu

References

[1] Rivest RL, Shamir A, Adleman L. A method for obtaining digital signatures and public-key cryptosystems. Communications of the ACM. 1978; **21**(2):120-126

[2] Dworkin MJ. SHA-3 Standard: Permutation-Based Hash and Extendable-Output Functions. Federal Inf. Process. Stds. (NIST FIPS) - 202. NIST Pubs; 2015. DOI: 10.6028/NIST.FIPS.202. Available from: https://www.nist.gov/publications/sha-3-standard-permutation-based-hash-and-extendable-output-functions?pub_id=919061

[3] Barker E. Recommendation for Key Management Part 1: General. National Institute of Standards and Technology. 2016. Report No.: NIST SP 800-57pt1r4. Available from: https://nvlpubs.nist.gov/nistpubs/SpecialPublications/NIST.SP.800-57pt1r4.pdf

[4] Bellovin SM. Frank Miller: Inventor of the one-time pad. Cryptologia. 2011; **35**(3):203-222

[5] Abushgra AA. Variations of QKD protocols based on conventional system measurements: A literature review. Cryptography. 2022;**6**(1):12

[6] Kato K. Non-orthogonality measure for a collection of pure quantum states. Entropy. 2022;**24**(5):581

[7] Bennett CH, Brassard G. An update on quantum cryptography. In: Workshop on the Theory and Application of Cryptographic Techniques. Berlin, Heidelberg: Springer; Aug 1984. pp. 475-480

[8] Gamel O. Entangled Bloch spheres: Bloch matrix and two-qubit state space. Physical Review A. 2016;**93**(6): 062320

[9] Abushgra A, Elleithy K. A shared secret key initiated by EPR authentication and qubit transmission channels. IEEE Access. 2017;**5**: 17753-17763

[10] Abushgra A. A New QKD Protocol Based upon Authentication by EPR Entanglement State. US: University of Bridgeport; 2018

[11] Einstein A, Podolsky B, Rosen N. Can quantum-mechanical description of physical reality be considered complete? Physics Review. 1935;**47**(10):777

[12] Gisin N, Ribordy G, Zbinden H, Stucki D, Brunner N, Scarani V. Towards practical and fast quantum cryptography. ArXiv Prepr Quant-Ph0411022. Quantum Physics. Cornell University; 2004. Available from: https://arxiv.org/abs/quant-ph/0411022 [Preprint]

[13] Serna EH. Quantum key distribution from a random seed. ArXiv Prepr ArXiv13111582. Quantum Physics. Cornell University; 2013. Available from: https://arxiv.org/abs/1311.1582

[14] Serna EH. Quantum key distribution protocol with private-public key. ArXiv Prepr ArXiv09082146. Quantum Physics. Cornell University; 2009. Available from: https://arxiv.org/abs/0908.2146

[15] Khan MM, Murphy M, Beige A. High error-rate quantum key distribution for long-distance communication. New Journal of Physics. 2009;**11**(6):063043

Chapter 5

Vulnerability of Quantum Information Systems to Collective Manipulation

Fernando Javier Gómez-Ruiz, Ferney Rodríguez, Luis Quiroga and Neil F. Johnson

Abstract

The highly specialist terms 'quantum computing' and 'quantum information', together with the broader term 'quantum technologies', now appear regularly in the mainstream media. While this is undoubtedly highly exciting for physicists and investors alike, a key question for society concerns such systems' vulnerabilities – and in particular, their vulnerability to collective manipulation. Here we present and discuss a new form of vulnerability in such systems, that we have identified based on detailed many-body quantum mechanical calculations. The impact of this new vulnerability is that groups of adversaries can maximally disrupt these systems' global quantum state which will then jeopardize their quantum functionality. It will be almost impossible to detect these attacks since they do not change the Hamiltonian and the purity remains the same; they do not entail any real-time communication between the attackers; and they can last less than a second. We also argue that there can be an implicit amplification of such attacks because of the statistical character of modern non-state actor groups. A countermeasure could be to embed future quantum technologies within redundant classical networks. We purposely structure the discussion in this chapter so that the first sections are self-contained and can be read by non-specialists.

Keywords: quantum communication, quantum computing, dicke model, radiation-matter interaction, loschmidt amplitude

1. Introduction

Technology is rapidly moving toward an era that will fully embrace the true "spookiness" (e.g. action-at-a-distance) of quantum mechanics, where quantum mechanical information processing systems will offer unique advantages over current classical counterparts [1–12]. Enormous investments have been made recently in quantum information technologies: in the United States with the NSF 'Quantum Leap' initiative [13], in Europe [14] and in China [15]. Among other global-scale applications, the idea of a quantum Internet is gaining significant traction [12, 16–21]. A key desirable feature of such a quantum Internet is that it provides enhanced security and novel functionalities not present in the existing classical Internet infrastructure. Human nature, however, will likely remain unchanged [12]. Just as human ingenuity can be used for good to drive quantum technology – which is the overriding narrative of most funding agencies [12, 13] – it could also be used for bad [22–29]. An obvious yet largely unanswered question then arises, of what might go wrong at scale in such a quantum system in terms of bad-actor manipulation?

Based on the recent successes of building and manipulating matter in the form of nanostructure systems, and light in the form of photon fields as carriers of information, it seems likely that a future quantum information processor or quantum Internet will comprise a matter-light system. In other words, it could easily end up containing N qubits (two-level quantum systems) made of matter that sit in some photonic environment (i.e. light) and hence bosonic field as in **Figure 1a**.

We hence focus our discussion in this chapter around this generic system setup in **Figure 1a**. As we will go on to explain in this chapter, our interest lies in seeing what the effect would be if N adversaries – who each choose one of the N qubits to attack – simultaneously manipulate the coupling λ between their chosen qubit and one of the bosonic modes by ramping λ up and down as in **Figure 1a**. The resulting pulse profile $\lambda(t)$ applied simultaneously between all N qubits and the bosonic mode is what we mean by collective manipulation of the quantum system by a hostile group, and the extent of the damage (i.e. change) that they cause to the system's initial overall quantum state will act as an indicator of the system's vulnerability to that type of collective attack.

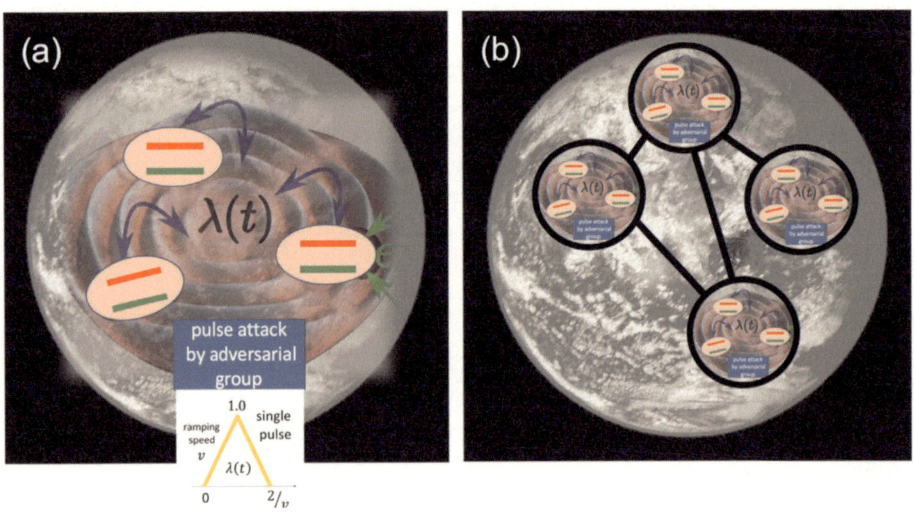

Figure 1.
Future quantum technologies. (a) Ultimate quantum technology limit of an extended geographical space covered by a quantum cloud in which the global quantum state is coherent [12]. This could be a spatially extended cavity of bosonic modes containing an arbitrary number N of qubits (i.e. two-level systems). A group of adversaries attack by applying a pulse interaction $\lambda(t)$ (see Eq. 6) between the qubits and a bosonic mode at speed v, with a duration $2/v$. Our conclusions are insensitive to the precise shape of the pulse, but we show a triangular pulse here for concreteness. (b) an intermediate, simpler version that could be built sooner than (a) because of lower technological demand. It features smaller versions of the $N = 3$ qubit-cavity system in (a), which are then interconnected through separate quantum or classical communication channels.

To provide some practical background, there is a long history of theoretical and empirical research exploring the properties of individual quantum nanoscale systems as potential qubits, e.g. a single superconductor nanostructure, or a single trapped ion, or a single semiconductor quantum dot [30]. The semiconductor quantum dot example, though perhaps less well-known, is essentially identical to the others in that it is a nanoscale system which has (at least) two well-defined discrete energy levels, each with their own wavefunctions, and hence has been termed an 'artificial atom' in the literature. The focus has been on the properties of the ground and excited states and how the quantum information in this system could be manipulated by external magnetic or electric fields, or by some particular photonic input [31]. The goal of most such studies has been to explore how the system can be switched from one eigenstate to another, as well as creating quantum superpositions of such states. There is also a lot of work on pairs of qubits (e.g. two superconductor nanostructures, two trapped ions, or two semiconductor quantum dots). This is because such a pair ($N = 2$) represents the smallest example of a system where quantum entanglement can be explored in spatially separate systems. The goal in that case is to understand better how to exploit the 'spooky action at a distance' property of the quantum world, in order to store and transfer quantum information and develop quantum gates as part of some larger scale quantum information processor, quantum computer or quantum network. These studies can be challenging both experimentally and theoretically, hence the question of what might happen with a larger number of qubits (e.g. $N = 3, 4 \ldots$ as in **Figure 1a**) has received far less attention. Yet recent research [32] inspired by reports of novel quantum effects in photosynthesis, has shown that new quantum phenomena can arise in systems with $N = 3, 4 \ldots$ qubits that do not occur in systems of $N = 1$ or 2 qubits. Given that exotic quantum many-body states like superconductivity are known to exist for $N \to \infty$, this suggests that new types of quantum information phenomena can also arise for $N = 3, 4, \ldots$ qubits that do not occur for $N = 1$ or 2 qubits.

This helps motivate the work in this present chapter, i.e. to analyze the potential vulnerabilities of systems with $N = 3, 4, \ldots$ etc. qubits in a bosonic environment. Any future quantum information processor or computer will of course contain $N \geq 3$ qubits in total, hence the importance of understanding such a system. In short, one might expect a quantum version of the phrase 'two's company but three is a crowd' and hence the vulnerabilities of a (quantum) crowd of qubits in a bosonic mode (i.e. $N \geq 3$) will be different from those of an individual qubit or a pair of qubits in a bosonic mode.

Heightening this need to understand $N \geq 3$ qubit-plus-bosonic-mode vulnerabilities from the quantum hardware side, is the fact that the study of attacks on quantum information systems has tended to focus around scenarios involving the exchange of information (qubit states) between two entities (so-called Alice and Bob). The third entity is simply a single adversary (Eve, the eavesdropper). While such Alice-Bob and Eve thought experiments are essential for understanding what happens to one or two qubits and their information in terms of cryptography and secrecy, there is therefore a gap in understanding the vulnerabilities of systems containing $N = 3$ or more qubits. Finally, since the Hilbert space of quantum states explodes in size as N increases, as does its internal complexity, the properties of an entangled state for $N = 3, 4 \ldots$ qubits-plus-bosonic mode can have entirely new opportunities for storing and processing quantum information. Hence the $N = 3, 4 \ldots$ qubits-plus-bosonic mode quantum state will likely have entirely new vulnerabilities to attack by adversaries, as we indeed show in this chapter.

In the social science field of conflict studies, this need to understand the effects of groups of adversaries is already well recognized, since this is how most insurgent and terrorist attacks occur [22, 24–28, 33] – with the added modern twist (which also applies to quantum information attacks in **Figure 1a**) that they need not all be in the same geographical location since they just need a means of synchronizing their actions (e.g. through some common clock) [12, 22, 24]. In other words, they all initiate the same manipulation $\lambda(t)$ between their chosen qubit and the bosonic mode in **Figure 1a** starting at exactly the same time.

In summary, this need to improve the understanding of vulnerabilities in $N \geq 3$ qubit-bosonic-mode quantum information systems explains the need for this chapter [12]. In particular, we will consider the following urgent question: *Can new types of vulnerabilities arise in future matter-light quantum technologies which will undoubtedly have $N \geq 3$ qubits?* We will show that the answer is a definitive 'yes': specifically, we will show that an entirely new form of threat arises by which a group of quantum-enabled adversaries can maximally disrupt the global quantum state of systems such as **Figure 1a** containing $N \geq 3$ qubits, in a way that is practically impossible to detect – and which may even get amplified by the way that humans naturally cluster into adversarial groups [24, 26–28, 34]. We offer a possible countermeasure but stress that there may be entire classes of such threats for which, as yet, there is no underlying scientific theory or understanding. Our analysis leverages the fact that whatever the future quantum technology, the necessary intercommunication across geographical distances will likely rely on electromagnetic waves – and hence a bosonic field of photons. This reinforces our reason for basing our analysis around a model of a generic, global quantum system involving a bosonic field [35].

We note that although we ourselves are condensed matter physicists with a background in semiconductors, everything that we discuss concerning qubits in this chapter is general. The individual qubits could be of any form, as long as they each have two identifiable eigenstates and that transitions can be made between them by some external driving field. We expect that some of this book's audience will have more of a quantum computing background (and will more naturally consider qubits more from the perspective of superconducting nanostructures or trapped ions). But in principle the two states per qubit could be anything – including exotic topological states of some kind. As long as the system can be mapped to the same form as the Hamiltonian in Eq. 1, which many can if the appropriate approximations are made, then all our results will hold and our conclusions will be unchanged.

We also note that some of the content of this chapter appears in earlier working papers on the arXiv preprint server (e.g. Ref. [12] and to a far lesser extent Ref. [36]). However the current chapter is not published anywhere else: pieces were posted on arXiv simply to gather feedback from the quantum information and quantum computing communities. For clarity, we will reference the prior working paper Ref. [12] multiple times in this chapter, to indicate parts where the two overlap.

We now discuss the layout of this chapter – which is particularly important to establish because our goal is to make it accessible to the broadest possible readership without compromising the details necessary to prove our claims. To achieve this, we have sectioned the presentation accordingly. Specifically, we dedicate Section 2 to a general overview of the field and our results. This can be read as a self-contained chapter by a general audience, with the intention that it explains what we did to answer this question, how we did it, what we found, and what it means – but without requiring too much technical detail. Section 3 then re-analyzes the problem at a deeper level in terms of the underlying physics. It intends to provide a specialist discussion of the physics of generating entangled states in light-matter systems, and hence the significance of the results that we present. Hence we have aimed Section 3 at a physics audience who wants to see the nuts and bolts of the problem and the specific Hamiltonian that we solve numerically. Section 4 provides a detailed discussion of the results in the language of quantum many-body physics. Section 5 provides a statement of our main conclusions.

2. Non-specialist overview of our analysis and findings

This section aims to give a general overview of what we did, why we did it, what results we obtained, and why they are important. We intend it to be readable as a self-contained section without requiring too much technical knowledge. Hence we will be fairly liberal and lax in our use of everyday language and terminology to avoid technical jargon [12]. Any reader requiring more rigorous discussion and details is referred to the later Sections 3 and 4, which can again be read as their own self-contained (but now highly technical) article.

We now explain why we focus on adversaries varying the qubit-bosonic-mode interaction λ as their method of attack [12]. Much of the existing analysis of quantum attacks and security is done based on the assumption that so-called gates exist (i.e. specific logic operations are possible). But such gates have to be achieved through some kind of interaction. This means that any quantum adversarial group will need to introduce a term into the system's equation (i.e. Hamiltonian) which will, they might hope, operate in a way similar to some set of gate operations. $\lambda(t)$ plays this role in **Figure 1a** and Eq. (6). If they are looking for large disruption, the question that we address in this paper is then how bad this could be. It is basic science that perturbing (e.g. attacking) a system involves interacting with it, hence our approach to analyzing the behavior of the quantum system with the interaction included makes sense. Second, it is basic math that it is impossible to recover the unique original quantum state if the final quantum state is orthogonal to it. Just imagine 2 vectors chosen from M orthogonal vectors. Picking one (the final state) says nothing about the other (original state). It is impossible to deduce the second from the first. The major surprise from our work is that such an orthogonal state, and hence maximal disruption, can emerge from such adversarial attacks using a simple pulse interaction in qubit-bosonic-mode systems with $N \geq 3$ qubits (see **Figure 2**). This finding is only made possible because of our detailed numerical calculations of the quantum many-body out-of-equilibrium system.

Figure 1a shows arguably the ultimate limit of such quantum technology in which an extended geographical space serves as a quantum cloud within which the quantum state is kept coherent: specifically, a spatially extended cavity with bosonic modes that contains an arbitrary number N of qubits (two-level systems) [37–39]. Given the current experimental success in distributing entangled photons over large distances [40–42], photons will likely provide the quantum 'glue' that binds together large-scale future quantum technologies globally, including a quantum Internet-of-Things. To be fully quantum mechanical, the system size should be within the coherence length of non-local correlations. For recent medium- to long-range practical systems, we refer to Refs. [41, 43, 44]. An intermediate version akin to **Figure 1b** could be achieved more quickly, in which smaller versions of such quantum node systems are interconnected through separate quantum or classical communication channels. Though still extremely challenging, we note that quantum coherence within cavities has already been demonstrated experimentally in a wide range of laboratory systems including solid-state and quantum optics [45–47].

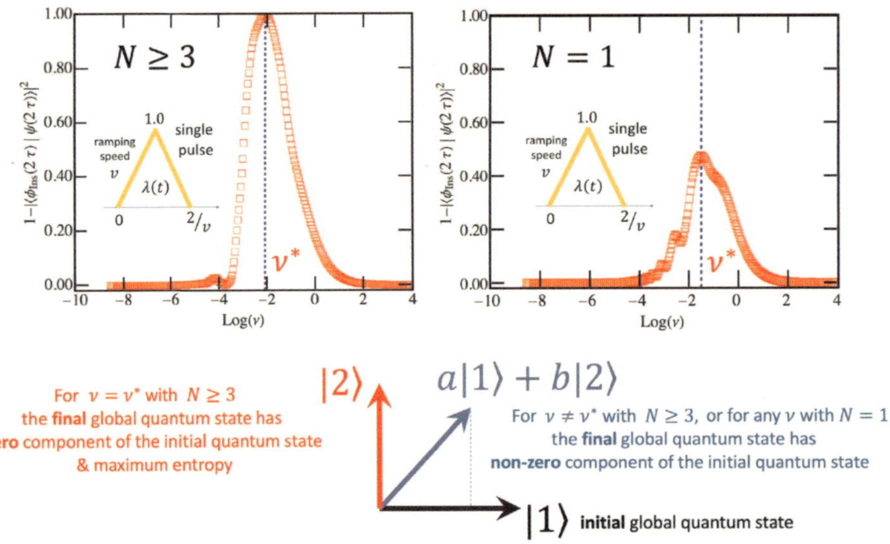

Figure 2.
*Disruption caused by an adversary. Top: Complete orthogonality to the initial global quantum state (which in practical terms translates to total disruption and hence maximal possible damage [12]) is produced by a hostile group (left panel) applying a pulse attack $\lambda(t)$ at pulse speed $v = v^*$ simultaneously to N qubits in a bosonic-mode as in **Figure 1a**. Since the final global quantum state for $v = v^*$ has no overlap with the initial one, the initial state cannot be filtered out from the final one – Hence maximum disruption. This is fundamentally different from an attack by a lone 'eve' (i.e. single adversary and hence $N = 1$) for whom there is no v that produces the same effect (right panel). The curve in the left panel is for group size $N = 3$ but is visually the same for any $N \geq 3$. We use an initial global quantum state of no photons here for simplicity, but our results can be generalized. Bottom: Schematic representation of this disruption.*

We now provide a brief description of our underlying theoretical model that provides our conclusions. This model is presented in detail in Section 3, while its output is discussed in more detail in Section 4. It is purposely chosen to be simple enough that it allows the development of quantitative results and intuition, and yet is realistic enough to capture the highly non-trivial empirical nature of quantum light-matter interactions [48, 49]. Since the N 'Eve' adversaries are ultimately humans or machines, we allow each of them to choose a single qubit from the N available: they are each going to then manipulate the interaction between their chosen qubit and the bosonic mode using a pulse-shaped time-dependent interaction form as in **Figure 1a**. Strictly speaking, our results are unchanged irrespective of whether each of these interactions is controlled by a single human, a set of humans, another machine, an algorithm, or a bot. We assume that each individual has modest capabilities: they can communicate classically once at some stage before the attack, concerning when to interact their qubit with the bosonic mode and with what temporal profile $\lambda(t)$. This classical communication can be achieved in many mundane ways, e.g. diffusion, announcement, or pass-it-on meaning that the N adversaries do not need to know each other or show any active coordination or collaboration during the subsequent pulse attack when $\lambda(t)$ is non-zero. The interaction pulse $\lambda(t)$ can have a very short duration, e.g. a fraction of a second is possible using current cavity technology [50–54]. Our results are similar for all up-down profile shapes and so we assume a triangular one for simplicity. We assume the resonant condition $\epsilon = \omega$ which has indeed been demonstrated in many laboratory systems experimentally [50–54]. Though we use a specific form of the system equation (i.e. Hamiltonian), similar overall conclusions should follow from many variants due to an established universal dynamical scaling [32, 35, 55–57] and the fact that they typically generate similar types of phase diagrams, and hence have similar collective states in the static λ limit [58–61]. Indeed, we have already shown elsewhere in the science literature that there is a universal dynamical scaling behavior for a particular class concerning their near-adiabatic behavior, in particular the Transverse-Field Ising model, the Dicke Model and the Lipkin-Meshkov-Glick model [35].

The disruption inflicted on the global quantum state by a group of $N = 3$ adversaries (since there are $N = 3$ qubits), utilizing a pulse interaction $\lambda(t)$, is vividly depicted in **Figure 2** (top left panel). This scenario stands in stark contrast to the outcome observed when dealing with a lone wolf attacker ($N = 1$ qubit), as illustrated in the top right panel of **Figure 2**. Remarkably, no conceivable choice of v for $N = 1$ can reproduce the disruptive effect caused by the collective action of $N = 3$ adversaries, nor can any number of asynchronous lone wolf attacks achieve a comparable outcome.

These findings hold true not only within the system depicted in **Figure 1a** but also within each individual quantum cavity delineated in **Figure 1b**. The consequential disorder, often quantified as "entropy", induced into the system is visually represented in **Figure 3**.

An intriguing observation emerges when examining the relationship between the pulse speed v and the manifestation of unique orthogonality for $N \geq 3$, as depicted in the left panel of **Figure 2**. It aligns remarkably closely, within numerical error, with the value at which the induced entropy reaches its maximum, as evidenced in **Figure 3**. This alignment provides compelling support for our central conclusion:

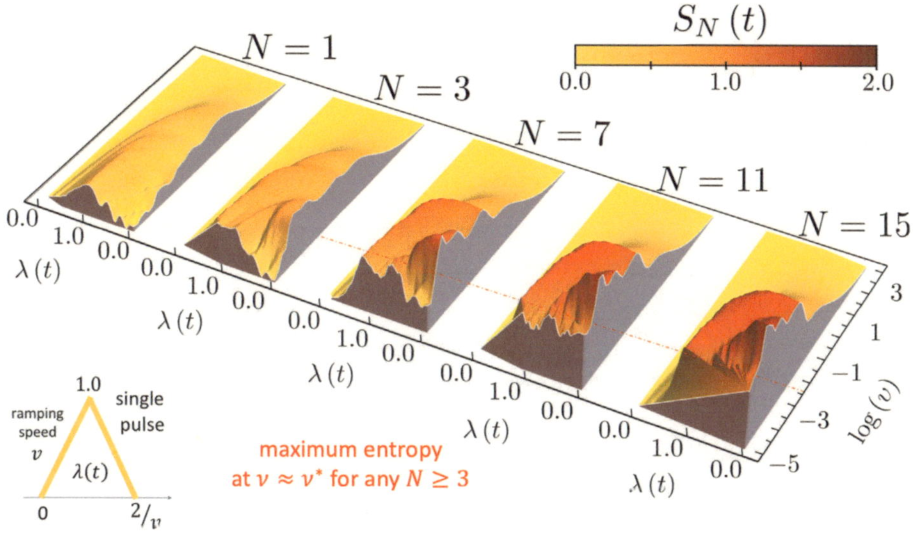

Figure 3.
Disorder (i.e. entropy) induced by attack. Von Neumann entropy S_N, which is a measure of the disorder in the quantum state, shown throughout the attack (i.e. as a function of $\lambda(t)$ and hence starting and ending at $\lambda(t) = 0$) for a given pulse speed v and a given number of qubits N. The number of attackers (adversaries) is the same as the number of qubits since each adversary chooses their own qubit. The general case $N \geq 3$ leads to a final global quantum state with maximum entropy when $v = v^*$ within numerical error, and hence pulse duration $2/v^*$. Our numerical calculations suggest that essentially the same result holds for all higher N, with the additional impact that the value of the maximum entropy (and hence the disruption) increases with N (i.e. peak becomes increasingly red in the figure).

groups comprising $N \geq 3$ adversaries pose a distinct and formidable disruptive threat to the integrity of future quantum systems.

Our main findings can be summarized as follows. First, adversarial 'Eve' groups employing intermediate attack speeds v^* will be able to generate maximal global quantum state disruption in systems containing $N \geq 3$ qubits – which means in practice *all* future quantum technology systems since none are likely to only have 1 or 2 qubits alone. Specifically, the final global quantum state contains zero component of the initial one and the corresponding Von Neumann entropy (which is indicative of the amount of disorder generated in the initial quantum state) is also a maximum within numerical uncertainty (**Figures 2** and **3**). No amount of filtering can then recover the initial global quantum state. Second, within numerical uncertainty, the speed at which this maximal disruption is obtained v^* is insensitive to N for $N \geq 3$, meaning that a general strategy exists that attackers can easily copy (i.e. plug-n-play) without needing to generalize or understand the underlying many-body quantum mechanics. Third, the attack requires no real-time communication, collaboration, or cooperation between any of the adversaries. They just need to agree ahead of time when to start the pulse, its maximum value $\lambda = 1$, and the speed v and hence duration $2/v$ of the pulse. From then on, they can operate independently as clock-synchronized individuals from any geographical location within one of the cavities in **Figure 1**.

Fourth, this attack leaves no clues that it has occurred based on the Hamiltonian (i.e. the Hamiltonian that we show later in Eq. (1) is unchanged after the pulse since $\lambda(t) = 0$ before and after) or based on the purity of the corrupted quantum state (i.e. quantum state purity remains one as long as there is no decoherence) and hence the attack can be practically impossible to trace. Fifth, the attack can be over within a fraction of a second since interactions between material qubits and boson cavity modes can be switched on and off very quickly using current technology [52–54]. Sixth, the impact of the attack (**Figure 4**) as measured in terms of an open-system generalization of the entropy from **Figure 3**, appears to become more robust to natural decoherence as the size of the quantum information processing system or quantum computer N increases further beyond 3 to higher numbers (i.e. higher N).

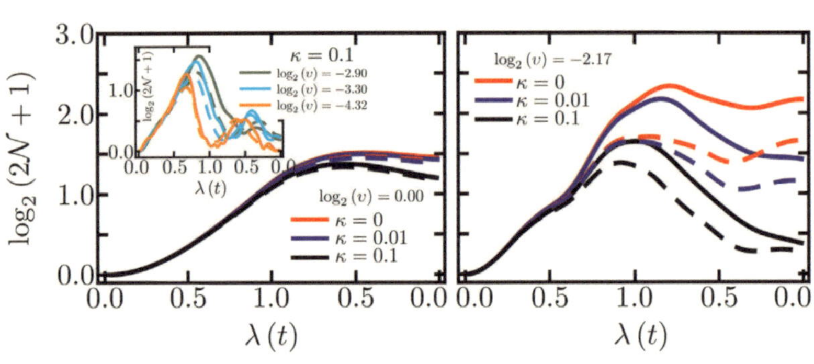

Figure 4.
Robustness to decoherence. The quantum logarithmic negativity (which can loosely be thought of as reflecting the amount of quantum entanglement) for an open version of our system (vertical axis) as a function of time during the pulse attack, and hence as a function of $\lambda(t)$. The quantum logarithmic negativity has become a widely accepted entanglement measure for an open system and incorporates the effects of natural decoherence and losses from the cavity. Results are shown for adversarial attacks on systems of $N = 5$ (solid lines) and $N = 11$ (dashed lines) qubits and for several values of decoherence κ. The results show a surprising robustness against decoherence and losses that increase with the number of qubits N in the quantum information processing system or quantum computer.

In stark contrast, lone 'Eve' attackers with no synchronization of their pulses, will achieve a maximum disruption that is not only smaller (**Figure 2**, top right panel) but which is fundamentally different in that the final global quantum state at $\upsilon = \upsilon^*$ still contains a finite amplitude of the initial state, and hence the correct state (i.e. initial state) can in principle be recovered using purification or distillation schemes. The perfect disruption produced by a synchronized attack on $N \geq 3$ qubits could not strictly be achieved by any number of lone attackers, even if they each attack many times since the overlap (**Figure 2**, top right panel) will always be non-zero. If we were to empower a single lone attacker with an extraordinary potential impact through a much larger λ, then it is in principle possible – but for $\lambda = 1$ we have shown that there is a fundamental difference between the two scenarios. This qualitative difference is also in complete contrast to classical conflict theory [62] where the disruption caused by adversarial groups of size N is assumed to simply scale as N, N^2 or at most N^δ $(\delta > 2)$ for any N.

Our final finding concerns the scenario in which the N qubits happen to be clustered together in some way, and hence the N adversaries that operate them will also be clustered together in the same way during the pulse attack. It turns out that such clustering of adversaries is actually a preferred pattern for operation of groups of adversaries – and remarkably, so too it has been shown experimentally that such clustering of qubits can give an enhanced coherence in the many-body quantum state – hence this coincidence could enhance the impact of such attacks. Specifically, previous work [24, 26–28, 34] has shown that modern adversarial groups tend to show a common power-law-like cluster distribution with exponent near 2.5 both offline and online [24, 26–28, 34], and hence that this is a natural pattern for how humans self-organize for adversarial activities. Such clustering might at first be expected to degrade the post-attack quantum state, akin to how islands of impurities might be expected to enhance single-particle scattering, and hence weaken the finding that the final state post-attack is orthogonal to the initial one. However, it has been shown experimentally [63] that the coherence of a global many-body quantum state is actually *strengthened* if there are matter clusters with an approximate 2.5 power-law size distribution [63, 64]. Taking the N adversaries as similarly clustered and allowing these clusters to generate a constant low-level background interaction with the quantum cloud – like the impurity clusters in the experiment in Ref. [63] – then the coherence and hence orthogonality of the final state should be similarly *strengthened* by any such adversarial clustering.

Countering this threat properly will require a new understanding of time-dependent quantum correlations in many-body light-matter systems. In the meantime, we note that the technology in **Figure 1b** will likely arrive before that in **Figure 1a**, meaning that a network architecture of connected cavities will be built first. In such a scenario, it would be possible to build redundancy into the network: then even if a perfectly orthogonal final state is generated within one cavity by a adversarial group attack at speed υ^*, it would be unlikely that this would be achieved simultaneously within a separate cavity, and so error correction schemes could be carried out across all cavities where each cavity is now a network node, and each node is crudely treated as a two-level system.

3. Detailed physics behind our calculations

This section analyzes in depth the calculations that we perform and the results that we obtain. It can be read as a stand-alone technical physics report, together with the detailed results in Section 4.

We first present a deeper discussion of the relevance of light-matter systems as our focus of attention. In the implementation of quantum communication infrastructures, diverse strategies have been employed. One approach involves setting up ad-hoc dark fiber networks designed solely for quantum communications. Another method integrates cutting-edge quantum technologies, such as Quantum Key Distribution (QKD) devices and fiber optics-based quantum links, into existing operational communication networks. This integration allows for the coexistence of quantum and classical links within the same infrastructure. The utilization of ad-hoc dark fiber networks ensures the exclusive use of these dedicated channels for quantum communications, minimizing interference and enhancing the overall security of quantum information transfer. On the other hand, the integration of quantum technologies into existing communication networks represents a more versatile approach, capitalizing on the infrastructure already in place. This strategy facilitates a seamless incorporation of quantum capabilities while sharing resources with traditional communication links. By exploring these diverse strategies, the implementation of quantum communication infrastructures can be tailored to specific needs, balancing considerations of security, efficiency, and adaptability within the broader landscape of communication networks.

Light has long been regarded as the most crucial component of information transmission, particularly since the advent of relativity. Photons may travel practically inconceivable distances, and they rarely interact among themselves unless matter particles in different environments mediate that interaction. Particularly, quantum information processing requires significant matter-light interactions. The idea of implementing selected photon interactions in the quantum domain has attracted a lot of attention lately. In terms of interacting photons, the development of active quantum matter seems reasonable involving crystalline states of light or even topological phases.

3.1 Temporal coupling $\lambda(t)$ from driving field

Now we discuss the light-matter system and hence justify our use of a time-dependent $\lambda(t)$ in our model Hamiltonian (see later Eq. 6) in order to represent the pulse-driven light-matter system – and hence why $\lambda(t)$ is the weapon used by each of the adversaries on their chosen qubit (N qubits total). For quantum systems embedded in complex environments, where extra degrees of freedom modulate the interaction between the quantum system of interest and a large reservoir, effective non-Markovian behaviors in the quantum system dynamics arise even though the reservoir itself can be described within a Markovian approximation. Consequently, although the phase imprinted by the excitation laser is indeed lost during the first steps of electron-exciton relaxation from the high-energy sector, this is not a sufficient reason to exclude any coherent-like behavior in the relaxing dynamics. Indeed it can be shown that for a wide class of phase-mixed states of the pump modes, results for the signal population can be obtained that are identical to those for a coherent population of those modes. To clarify this critical point, we now show that our basic premise is justified for a variety of reasons. According to the extensive literature concerning light-matter Hamiltonians, in the classical limit, the system can be considered equivalent to two coupled harmonic oscillators. This information is enough to gain analytical insight into the solution of the resulting quadratic system. The driven system in this limit is described by two coupled harmonic oscillators with a time-dependent coupling frequency. Consequently, for this purpose, we will consider a simplified model for the parametric process that contains just 3 boson modes (for the sake of simplicity we describe now the N dimer subsystem in the low excitation limit as an effective boson b mode), as described by the Hamiltonian:

$$\hat{H} = \omega_a \hat{a}^\dagger \hat{a} + \chi\left(\hat{a}^\dagger \hat{a}\right)^2 + \omega_b \hat{b}^\dagger \hat{b} + \omega_c \hat{c}^\dagger \hat{c} + g\left(\hat{a}^\dagger \hat{b}^\dagger \hat{c}^2 + \hat{a}\hat{b}\hat{c}^{\dagger 2}\right) \tag{1}$$

where the operators $\hat{a}$, $\hat{b}$ and $\hat{c}$ correspond to bosonic modes (e.g. vibronic, exciton, and high energy exciton modes) where we allow for anharmonic terms of strength χ. We now consider the effect of the pump state on the dynamics of this simple, but representative, model. In particular, we consider the excitation of high-energy electron states, which indirectly feeds (through the relaxation process) an effective pump reservoir that follows the applied radiation pulse shape. We assume that $\left(\hat{a}^\dagger \hat{b}^\dagger \hat{c}^2 + \hat{a}\hat{b}\hat{c}^{\dagger 2}\right) = h(t)\left(\hat{a}^\dagger \hat{b}^\dagger + \hat{a}\hat{b}\right)$ where $h(t)$ represents the applied pulse shape. It is usually argued that the expectation value $\langle \hat{c}^{\dagger 2} \rangle$ ($\langle \hat{c}^2 \rangle$) is different from zero only if the high energy reservoir states have coherent populations. Since the laser pulse excites electrons at a higher energy, the excess energy might be expected to relax into the exciton region giving rise to a coherent interaction. However this is not necessarily the case: after non-resonant excitation, the phase imprinted by the excitation laser is generally lost. The appearance of a well-defined phase is often regarded as the true characteristic feature of a coherent state. However, a careful analysis of unitary dynamics from mixed states, such as those produced by incoherent relaxation processes, shows that coherent-like behaviors can often be obtained. To justify this last claim we compute the time evolution of observables under two kinds of pump initial states: (i) A pure initial state like $|\Psi\rangle = |0_a\rangle|0_b\rangle|\alpha_c\rangle$, denoting the vacuum state for both modes and a pump coherent state. (ii) A statistical mixed state with no phase information at all, given by a density matrix $\hat{\rho}_P = \int_0^{2\pi} d\theta P(\theta) \hat{\Pi}_p(\theta)|\Psi\rangle\langle\Psi|\hat{\Pi}_p^{-1}(\theta)$, where $\hat{\Pi}_p(\theta) = e^{i\hat{N}\theta}$, with $\hat{N} = \hat{a}^\dagger \hat{a} + \hat{b}^\dagger \hat{b} + \hat{c}^\dagger \hat{c}$, denotes a phase smearing operator, given the fact that it takes a pump coherent state $|\alpha_c\rangle$ to a different phase coherent state $|e^{i\theta}\alpha_c\rangle$. The function $P(\theta)$ fixes the pump phase smearing effect with $P(\theta) \geq 0$ and $\int_0^{2\pi} d\theta P(\theta) = 1$. Since $\left[\hat{H}, \hat{N}\right] = 0$, it follows that the time-evolution operator $\hat{U}(t) = e^{-i\hat{H}t}$ commutes with the phase smearing operator $\hat{\Pi}_p(\theta)$. It is now an easy task to obtain for any observable like $\hat{a}^{\dagger k}\hat{a}^l$, the time-evolution as

$$\begin{aligned} \left\langle \hat{a}^{\dagger k}\hat{a}^l \right\rangle_P &= Tr\left\{\hat{a}^{\dagger k}\hat{a}^l \hat{\rho}_P(t)\right\} \\ &= \int_0^{2\pi} d\theta P(\theta) Tr\left\{\hat{U}^{-1}(t)\hat{\Pi}_p^{-1}(\theta)\hat{a}^{\dagger k}\hat{a}^l \hat{\Pi}_p(\theta)\hat{U}(t)|\Psi\rangle\langle\Psi|\right\}. \end{aligned} \tag{2}$$

Since $\hat{\Pi}_p^{-1}(\theta)\hat{a}^{\dagger k}\hat{\Pi}_p(\theta) = e^{-ik\theta}\hat{a}^{\dagger k}$ and $\hat{\Pi}_p^{-1}(\theta)\hat{a}^l\hat{\Pi}_p(\theta) = e^{il\theta}\hat{a}^l$ it follows that

$$\left\langle \hat{a}^{\dagger k}\hat{a}^l \right\rangle_P = \left\langle \hat{a}^{\dagger k}\hat{a}^l \right\rangle_0 \int_0^{2\pi} d\theta P(\theta) e^{-i(k-l)\theta} \tag{3}$$

where $\left\langle \hat{a}^{\dagger k}\hat{a}^l \right\rangle_0$ corresponds to the initial state with the pump in a coherent state. It is evident that the population dynamics of the subsystem ($k = l = 1$) is fully insensitive to this class of phase smearing in the pump state, $\langle \hat{a}^\dagger \hat{a} \rangle_P = \langle \hat{a}^\dagger \hat{a} \rangle_0$, as well as other vibrational correlations as long as the pump phase smearing probability $P(\theta)$ remains practically constant. The main physical ingredients of a general, complex light-matter system can therefore be captured by this simple 3-mode Hamiltonian. Therefore we can conclude that for a wide class of coherent pump-plus-relaxation process conditions, our main results are indeed meaningful. Hence the replacement of $\hat{c}$-pump operators by complex numbers – which consequently yields a time-dependent coupling strength $\lambda(t)$ – is justified. Also, the range of validity of our assumption is the same as the usual one for the parametric approximation which requires a highly populated coherent state, $|\alpha_c| \gg 1$, and short times, $gt \ll 1$. These conditions are precisely identical to those under which our model fits with previous studies of coherence generation: high excitation and rapid relaxation dynamics. Therefore, there is indeed a formal justification for reducing the last terms to $gh(t)\left(\hat{a}^\dagger \hat{b}^\dagger + \hat{a}\hat{b}\right)$ where $h(t)$ represents the applied pulse shape – hence justifying the time-dependent interaction $\lambda(t) \sim gh(t)$.

Next, we discuss the fact that any incident electromagnetic (light) field $\vec{E}$ generates an internal polarization field $\vec{P}$ within the material, given exactly by Maxwell's Equations. The equation describing the time-domain behavior in a general, anisotropic and nonlinear medium subject to a general time and position-dependent light field $\vec{E}$ is given by:

$$\nabla \times \nabla \times \vec{E} + \mu_0 \sigma \frac{\partial \vec{E}}{\partial t} + \mu_0 \frac{\partial^2 \overrightarrow{\overrightarrow{\epsilon}} \cdot \vec{E}}{\partial^2 t} = -\mu_0 \frac{\partial^2 \vec{P}}{\partial^2 t} \tag{4}$$

in which the standard symbols have their well-known meaning from electromagnetic theory (e.g. $\overrightarrow{\overrightarrow{\varepsilon}}$ is a complex second-order tensor). If the medium is lossless then $\sigma = 0$ and so this equation can be rewritten as:

$$\left[\nabla \times (\nabla \times) + \frac{1}{\epsilon_0 c^2} \frac{\partial^2}{\partial t^2} \overrightarrow{\overrightarrow{\epsilon}} \cdot \right] \vec{E} = -\frac{1}{\epsilon_0 c^2} \frac{\partial^2 \vec{P}}{\partial^2 t}. \tag{5}$$

Though nonlinear and anisotropic in general, the presence of $\partial^2/\partial t^2$ terms for $\vec{E}$ and $\vec{P}$ in both equations means that a pulse in $\vec{E}$ will generate a similar pulse in $\vec{P}$, and hence a pulse in the internal electric field dynamics coupling the electronic and vibrational systems (i.e. a pulse in $\lambda(t)$).

Our focus is on near resonant conditions since these are the most favorable for generating large coherences. We only consider one such resonance for simplicity, however, this can be generalized by matching up different excitation energies ϵ', ϵ'', etc. to the nearest vibrational energies ω', ω'' etc., and then solving the Hamiltonian in the same way for each subset (ε', ω') etc. For example, if the N components are partitioned into n subpopulations, where each subpopulation has the same resonant energy and vibrational mode but where these values differ between subpopulations, the total Hamiltonian will approximately decouple into $H^{(1)} \oplus H^{(2)} \oplus H^{(3)} \ldots \oplus H^{(N)}$. Any residual coupling between these subpopulations can then be treated as noise.

3.2 Details of our hamiltonian model

This section provides a detailed discussion of our model using the specialist language of many-body physics and quantum information.

Quantum computers contain approximate two-level systems, where the difference between the two levels is the excitation energy, that are invariably coupled to other bosonic subsystems (e.g. light) as in **Figure 1**. Each adversary can perturb one such qubit system. We describe this using a generalized Dicke Model. Our model features a set of N identical two-level systems (commonly referred to as qubits) each of which is coupled to a single bosonic mode. It can be described by the following microscopic Hamiltonian:

$$\hat{H}(t) = \frac{\epsilon}{2}\sum_{i=1}^{N}\hat{\sigma}_z^i + \omega\hat{a}^\dagger\hat{a} + \frac{\lambda(t)}{\sqrt{N}}\left(\hat{a}^\dagger + \hat{a}\right)\sum_{i=1}^{N}\hat{\sigma}_x^i. \tag{6}$$

We purposely avoid common approximations such as the rotating-wave approximation. Here σ_α^i denotes the Pauli operators for qubit i $(\alpha = x, z)$; $\hat{a}^\dagger(\hat{a})$ is the creation (annihilation) operator of the radiation field. ϵ and ω represent the qubit and field transition frequencies respectively; and $\lambda(t)$ represents the strength of the radiation-matter interaction (i.e. it represents the adversaries' manipulation) at time t which can be varied over time. In many situations such as those we consider here in our work, the dynamics of the Dicke model do not require the consideration of the entire $2^{\otimes N}\otimes\mathbb{N}$ dimensional Hamiltonian. Instead, $SU(2)$ collective operators $\hat{J}_\alpha = \frac{1}{2}\sum_{i=1}^{N}\hat{\sigma}_\alpha^i$ can be used. The Hamiltonian can then be written in the following form:

$$\hat{H}(t) = \epsilon\hat{J}_z + \omega\hat{a}^\dagger\hat{a} + \frac{2\lambda(t)}{\sqrt{N}}\hat{J}_x\left(\hat{a}^\dagger + \hat{a}\right). \tag{7}$$

The static properties of the Dicke Model have been widely studied and characterized in the last two decades [65, 66]. It is well known that, in the thermodynamic limit $N \to \infty$, it exhibits a second-order quantum phase transition (QPT) [67] at $\lambda_c = \sqrt{\epsilon\omega}/2$ with order parameter $\hat{a}^\dagger\hat{a}/J$, separating the normal phase at $\lambda_c < \sqrt{\epsilon\omega}/2$ from the superradiant phase in which there is a finite value of the macroscopic order parameter, e.g. finite boson expectation number [35]. When the static coupling parameter is above this critical value λ_c, the ground state of the system is characterized by a non-zero expectation value of the excitation operators,

$$\langle\hat{N}_b\rangle \equiv \langle\hat{a}^\dagger\hat{a}\rangle \quad \text{and} \quad \langle\hat{N}_q\rangle \equiv \left\langle\hat{J}_z - \frac{N}{2}\right\rangle. \tag{8}$$

When $\lambda < \lambda_c$, the order parameters are zero. Because of this, the region when $\lambda > \lambda_c$ is called the ordered or superradiant phase, while the region when $\lambda < \lambda_c$ is called the normal phase. Near the phase-boundary in the vicinity of this superradiant phase, there is a dependence of the order parameter as follows: $\langle\hat{N}_b\rangle \propto (\lambda - \lambda_c)$ and $\langle\hat{N}_q\rangle \propto (\lambda - \lambda_c)^{1/2}$ [68]. This power law behavior is typical of second-order phase transitions where the critical exponents are characteristic of the universality class to which the model belongs. In the ordered quantum phase $(\lambda > \lambda_c)$, the $\mathbb{Z}_2$ symmetry related to parity is spontaneously broken, which originates from the fact that the thermodynamic limit ground state is two-fold degenerate corresponding to the two different eigenvalues of $\hat{P}$. Also at the QPT, the Dicke model presents an infinitely degenerate vanishing energy gap [68].

We use the Loschmidt Amplitude (LA), which is the overlap between the system's initial quantum state and its time-evolved state, as a new way to quantify quantum vulnerability: when the overlap is zero, it means orthogonality and hence maximum disruption has been created by the adversarial perturbation $\lambda(t)$. The logarithm of this LA value then presents a singularity. We note that the LA is one of the most significant quantities in modern quantum physics research. In particular, it has been used to detect time evolution singularities in quenched qubit-cavity system [69] and even for topological qubits inserted in photon cavities [70]. Results for the special case of rectangular pulses are depicted in **Figure 5**. Specifically, the LA defined as $\mathscr{L}(t) = \langle\Psi(0)|\Psi(t)\rangle$ represents the fidelity of the time-evolved agent state $|\Psi(t)\rangle$ with respect to the initially prepared state $|\Psi(0)\rangle$. When the logarithm of the LE value is close to zero, by this definition, there is a strong overlap between the initial state and the time-evolved state. Computationally, we perform a brief quench to the final state, assuming that the two-level systems are initially prepared in the ground state. The return probability of the time-evolved state to the initial state is the definition of the Loschmidt Echo (LE) L in the simplest of all quench protocols. The appearance of zeros in the Loschmidt Echo implies localization transitions, also known as dynamical phase transitions. It is worth noting that the longer time-evolved states of the post-quench Hamiltonian with time-dependent interactions differ significantly from the original state in which matter and light were created. As a result, the Loschmidt Amplitude (the scalar product of plane-wave and extended time-evolved states) vanishes at key points, indicating dynamical phase transitions. This means that the system's quantum state is now orthogonal and hence the disruption is maximal. For the special case where the excitation energy goes to zero, the LE could be obtained analytically as:

$$\mathscr{L}^N(t) = \frac{1}{2^{N-1}} \sum_{n=0}^{\left\lceil \frac{N}{2} \right\rceil} \binom{n}{N} e^{-\frac{\lambda_M^2}{N}\left(\frac{N}{2}-n\right)^2 \alpha(t)} \tag{9}$$

where

$$\alpha(t) = \frac{1}{3\omega_c^4 t^2}\left[9 - 12e^{-i\omega_c t} + 3e^{-2i\omega_c t} - 2i\omega_c t\left(3 - (\omega_c t)^2\right)\right]. \tag{10}$$

We leave the pursuit and analysis of such analytical forms to future work.

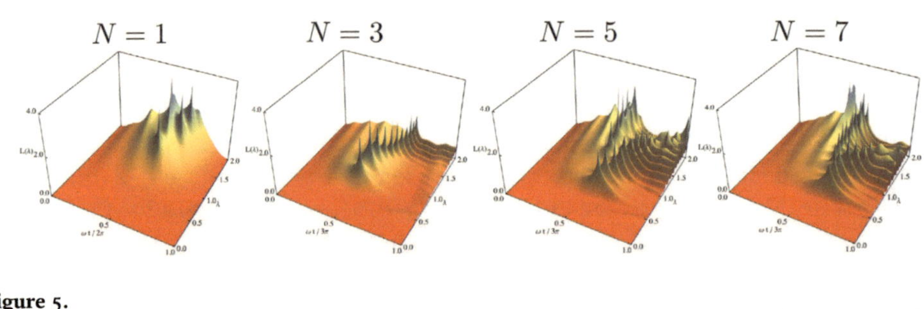

Figure 5.
Loschmidt Echo (LE) for a constant pulse λ. In the realm of deep strong coupling, a sequence of singularity peaks in the Loschmidt Echo becomes apparent, with their emergence dictated by both the coupling parameter λ and the time t following the onset of coupling.

4. Results

We find that the Loschmidt Echo (LE) shows kinks at important points as the system size N grows. At these points, the overlap with the original state is minimal, i.e. the overlap is near zero, hence the logarithm of the overlap shows a feature akin to a finite-size version of a singularity. The dynamics are such that over time, the state will spread in Hilbert space due to the coupling with confined light, reducing the peaks. Crucially, even while the light effectively functions as a dissipative information system, this expansion of cusps is broad and does not require specific adjustments of parameters. For a time-independent $\lambda/\omega_c < 1$, the overlap between two unique ground states in the Dicke model approaches zero as the number N of two-level systems rises. Because of this, both of the two ground states are mutually orthogonal in the thermodynamic limit. This effect is known as the Anderson orthogonality catastrophe and occurs in systems with infinitely many degrees of freedom when two physical states that correspond to two arbitrarily near sets of parameters (two arbitrarily comparable physical conditions) become orthogonal states of one another. This effect has previously been investigated in many-body physics. In contrast, a large peak develops in a band where the overlap approaches one for values of $\lambda/\omega_c > 1$. Surprisingly, we find that the system is susceptible to losing the original data for this collection of parameters. Thus, for zeros emerging at larger levels of $\lambda(t)$, we may observe that the composition of the initially prepared state becomes progressively dominated by the contribution of different two-level system states at higher energy states. An N-qubit circuit with an initial state for the unexcited agent and zero communication photons, with $N = 1,3,5,7,9$. **Figure 6** shows the LE results, where two unique zones are identified. (1) The LE travels to a single zone (vulnerable zone), and (2) a complicated structure makes the system immune to external perturbations. The two-dimensional representation of the complicated structure is shown in **Figure 6b,d,f,h**, on the right side. We have demonstrated that it is possible to achieve a regime where the light controls the zones in which the system fails to return to the initial state. This ability to manipulate the initial state so that it abruptly collapses into orthogonal states as a function of λ and t, represents a previously unknown threat to quantum technologies.

Our calculation of the impact of decoherence, as discussed in Section 2, uses the density matrix approach of Ref. [71] and a widely accepted measure of the open-system entanglement, the quantum negativity. We find that the presence of decoherence does not change our main conclusions. Not only do our main results survive well with increasing decoherence κ, the strength and robustness of the many-body coherence both increase with N. (For very large N, other system-level decoherence mechanisms may of course set in). Different temperatures are simulated by varying the average number of phonons $\overline{n}$, choosing values typical of the low temperatures in most experimental realizations, and including a thermal distribution in the initial density matrix [71]. Again our main conclusions are qualitatively unchanged.

Finally, we stress that our focus here has been on the question of collective coherence in any setting where there is some pulsed perturbation of the system due to adversaries manipulating the system. However the external profile $\vec{E}(t)$ and hence $\lambda(t)$ can be general. It need not be a pulse. Also the application of our system Hamiltonian analysis could be to transport experiments as well as optical experiments, or any combination of these.

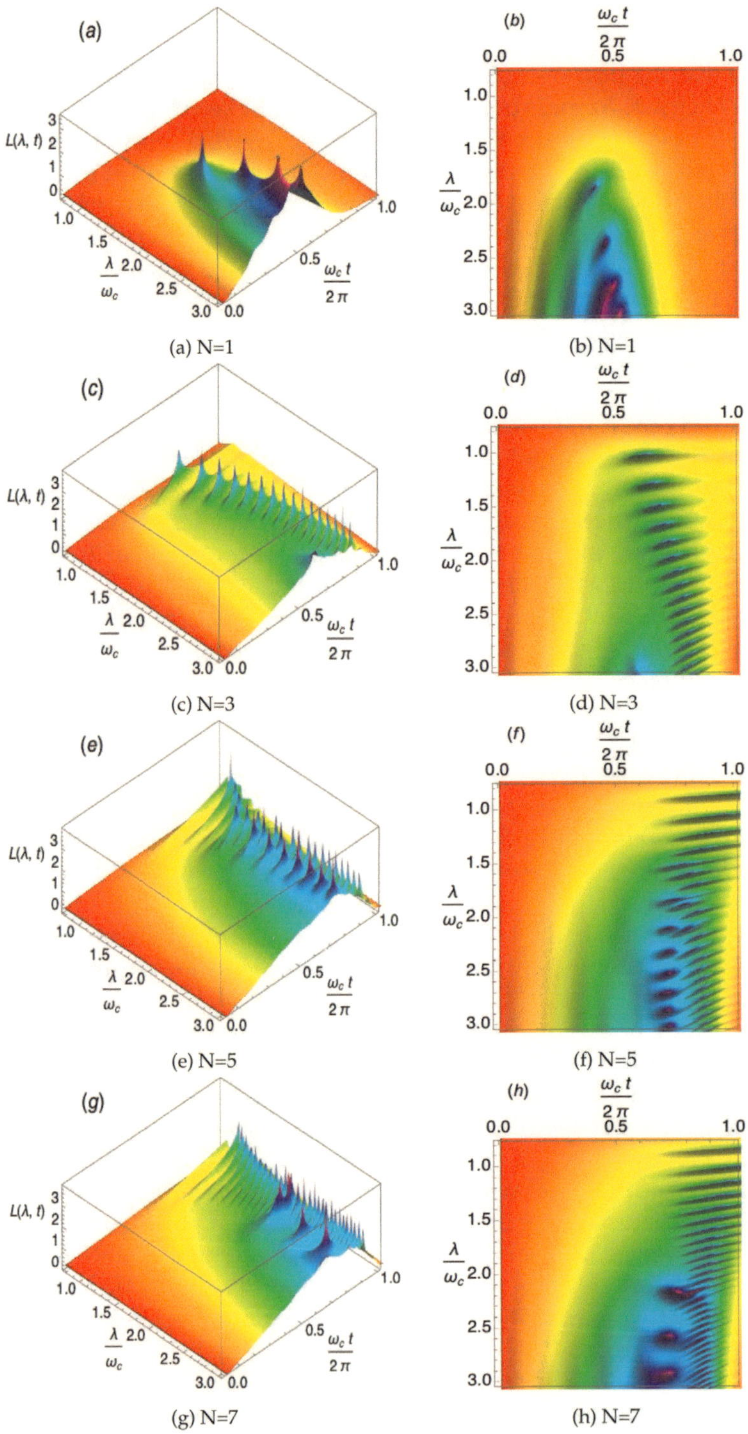

(a) N=1 (b) N=1

(c) N=3 (d) N=3

(e) N=5 (f) N=5

(g) N=7 (h) N=7

Figure 6.

Detailed output of our model Hamiltonian calculations for N = 1,3,5,7. The time-dependent interaction (i.e. adversaries' manipulation) is proportional to vt. For N ≥ 3, the singularities (and hence the disruption) appear increasingly strong as N increases.

5. Conclusions

We have addressed a key question for society concerning future quantum technologies: specifically, their vulnerability to collective manipulation. Specifically, we presented and discussed a new form of vulnerability in such systems that we identified based on detailed many-body quantum mechanical calculations. This new vulnerability will enable hostile groups to inflict maximal disruption on the global quantum state in such systems and hence will threaten their core functionality. These attacks will be practically impossible to detect since they introduce no change in the Hamiltonian and no loss of purity; they require no real-time communication; and they can be over within a second. We also predicted that such attacks will be amplified by the statistical character of modern non-state actor groups. A countermeasure could be to embed future quantum technologies within redundant classical networks.

We also analyzed more deeply the origins of this disruption by adversaries. Specifically, we showed in Section 3 that for a broad range of parameters, the system-level disruption (i.e. transitions) are characterized by curves in the appropriate Loschmidt functions at important points. By examining the numerical solution derived for the Loschmidt Amplitude, we found that the singularities were a component of a more intricate underlying structure in the system's temporal development. We used the Lodschmidt Echo to draw a connection between the macroscopic dynamics and the quantum vulnerability.

Acknowledgements

Spanish MCIN supported part of this research with funding from European Union Next Generation EU (PRTRC17.I1) and Consejeria de Educación from JCyL through the QCAYLE project, as well as MCIN projects PID2020-113406GB-I00 and RED2022-134301-T. F.J.R. and L.Q. are thankful for financial support from Facultad de Ciencias-UniAndes projects INV-2021-128-2292 and INV-2023-162-2833. N.F.J. is supported solely by U.S. Air Force Office of Scientific Research awards FA9550-20-1-0382 and FA9550-20-1-0383, as well as The John Templeton Foundation. The views and conclusions contained herein are solely those of the authors and do not represent official policies or endorsements by any of the entities named in this book chapter. An initial working paper which forms the basis for part of this paper, was posted as a preprint on the well-known arXiv preprint server: 'Quantum Terrorism: Collective Vulnerability of Global Quantum Systems' in 2019 https://arxiv.org/abs/1901.08873.

Author details

Fernando Javier Gómez-Ruiz[1], Ferney Rodríguez[2], Luis Quiroga[2] and Neil F. Johnson[3*]

1 Departamento de Física Teórica, Atómica y Óptica, Universidad de Valladolid, Valladolid, Spain

2 Departamento de Física, Universidad de los Andes, Bogotá D.C., Colombia

3 Physics Department, George Washington University, Washington D.C., USA

*Address all correspondence to: neiljohnson@gwu.edu

References

[1] IOP Publishing. See Special Set of "Quantum Horizons" Articles in Physics World. Available from: https://physicsworld.com/p/collections/quantum-horizons/

[2] Ekert A. Quantum cryptography based on Bell's theorem. Physical Review Letters. 1991;**67**:661

[3] Lloyd S. Ultimate physical limits to computation. Nature. 2017;**406**:1047

[4] Nickerson N, Li Y, Benjamin S. Topological quantum computing with a very noisy network and local error rates approaching one percent. Nature Communications. 2013;**4**:1756

[5] Ekert A, Renner R. Nature. 2014;**507**:443. DOI: 10.1038/nature13132

[6] Barz S, Kashefi E, Broadbent A, Fitzsimons JF, Zeilinger A, Walther P. Demonstration of blind quantum computing. Science. 2012;**335**(6066):303-308. Available from: http://science.sciencemag.org/content/335/6066/303

[7] Albash T, Lidar DA. Adiabatic quantum computation. Reviews of Modern Physics. 2018;**90**:015002. DOI: 10.1103/RevModPhys.90.015002

[8] Zurek WH. Quantum Darwinism, classical reality, and the randomness of quantum jumps. Physics Today. 2014;**67**(10):44. DOI: 10.1063/PT.3.2550

[9] Dziarmaga J, Zurek WH, Zwolak M. Non-local quantum superpositions of topological defects. Nature Physics. 2011;**8**:49. DOI: 10.1038/nphys2156

[10] Ma XS, Kofler J, Qarry A, Tetik N, Scheidl T, Ursin R, et al. Quantum erasure with causally disconnected choice. National Academy of Sciences of the United States of America. 2013;**110**(4):1221-1226. Available from: https://www.pnas.org/content/110/4/1221

[11] Kaku M. Quantum Supremacy. New York: Doubleday; 2023

[12] Johnson NF, Gómez-Ruiz FJ, Rodríguez FJ, Quiroga L. (Working Paper) Quantum terrorism. Arxiv. 2019. Available from: https://arxiv.org/abs/1901.08873

[13] National Science Foundation (NSF). NSF in Quantum Technologies at. Available from: gov/events/event_summ.jsp?cntn_id=296901

[14] Innovate UK KTN. European Commission in Quantum Technologies. Available from: https://qt.eu and https://ktn-uk.co.uk/interests/quantum-technologies

[15] National Natural Science Foundation of China. NSFC. Available from: https://www.nsfc.gov.cn

[16] Pirandola S, Braunstein SL. Physics: Unite to build a quantum internet. Nature. 2016;**532**:169. Available from: https://www.nature.com/news/physics-unite-to-build-a-quantum-internet-1.19716

[17] Castelvecchi D. The quantum internet has arrived (and it hasn't). Nature. 2018;**554**:289. Available from: https://www.nature.com/articles/d41586-018-01835-3#correction-0

[18] Wehner S, Elkouss D, Hanson R. Quantum internet: A vision for the road ahead. Science. 2018;**362**:eaam9288. Available from: http://science.

sciencemag.org/content/362/6412/
eaam9288

[19] Liao SK, Cai WQ, Handsteiner J, Liu B, Yin J, Zhang L, et al. Satellite-relayed intercontinental quantum network. Physical Review Letters. 2018; **120**:030501. DOI: 10.1103/PhysRevLett.120.030501

[20] Ren JG, Xu P, Yong HL, Zhang L, Liao SK, Yin J, et al. Ground-to-satellite quantum teleportation. Nature. 2017;**549**:70. DOI: 10.1038/nature23675

[21] Cartwright J. The Dream of a Quantum Internet Is Closer than you Might Think, Physics World. 2023 . Available from: https://physicsworld.com/a/the-dream-of-a-quantum-internet-is-closer-than-you-might-think/

[22] Schuurman B et al. End of the lone wolf: The typology that should not have been. Studies in Conflict & Terrorism. 2018;**0**(0):1-8. DOI: 10.1080/1057610X.2017.1419554

[23] Gill P et al. Terrorist use of the internet by the numbers quantifying Behaviors, patterns, and processes. Criminology and Public Policy. 2017; **16**(0):99

[24] Johnson N et al. New online ecology of adversarial aggregates: ISIS and beyond. Science. 2016;**352**(6292): 1459-1463

[25] Johnson NF et al. Pattern in escalations in insurgent and terrorist activity. Science. 2011;**333**:81

[26] Johnson NF et al. Simple mathematical law benchmarks human confrontations. Scientific Reports. 2013; **3**:3463. DOI: 10.1038/srep03463

[27] Bohorquez JC et al. Common ecology quantifies human insurgency. Nature. 2009;**462**:911

[28] Manrique PD et al. Generalized gelation theory describes onset of online extremist support. Physical Review Letters. 2018;**121**:048301. DOI: 10.1103/PhysRevLett.121.048301

[29] Johnson NF. To slow or not? Challenges in subsecond networks. Science. 2017;**355**(6327):801. Available from: http://science.sciencemag.org/content/355/6327/801

[30] Quiroga L, Ardila D, Johnson N. Spatial correlation of quantum dot electrons in a magnetic field. Solid State Communications. 1993;**86**:775

[31] Labonté L, Alibart O, D'Auria V, Doutre F, Etesse J, Sauder G, et al. Integrated photonics for quantum communications and metrology. PRX Quantum. 2024;**5**:010101. DOI: 10.1103/PRXQuantum.5.010101

[32] Gómez-Ruiz FJ, Acevedo OL, Rodríguez FJ, Quiroga L, Johnson NF. Energy transfer in N-component nanosystems enhanced by pulse-driven vibronic many-body entanglement. Scientific Reports. 2023;**13**:19790. DOI: 10.1038/s41598-023-46256-z

[33] Robb J. Brave New War. New York: Wiley; 2008

[34] Spagat M et al. Fundamental patterns and predictions of event size distributions in modern wars and terrorist campaigns. PLoS One. 2018; **13**(10):1-13. DOI: 10.1371/journal.pone.0204639

[35] Acevedo OL, Quiroga L, Rodríguez FJ, Johnson NF. New dynamical scaling universality for quantum networks across adiabatic

quantum phase transitions. Physical Review Letters. 2014;**112**:030403. DOI: 10.1103/PhysRevLett.112.030403

[36] Gómez-Ruiz FJ, Acevedo OL, Rodríguez FJ, Quiroga L, Johnson NF. (Working Paper) Pulsed Generation of Quantum Coherences and Non-Classicality in Light-Matter Systems. Arxiv. 2017. Available from: https://ar5iv.labs.arxiv.org/html/1706.07761

[37] Ritter S, Nölleke C, Hahn C, Reiserer A, Neuzner A, Uphoff M, et al. An elementary quantum network of single atoms in optical cavities. Nature. 2012;**484**:195. DOI: 10.1038/nature11023

[38] Mirza IM, Schotland JC. Two-photon entanglement in multiqubit bidirectional-waveguide QED. Physical Review A. 2016;**94**:012309. DOI: 10.1103/PhysRevA.94.012309

[39] Lloyd S. Enhanced sensitivity of photodetection via quantum illumination. Science. 2008;**321**(5895): 1463-1465. Available from: http://science.sciencemag.org/content/321/5895/1463

[40] Hofmann J, Krug M, Ortegel N, Gérard L, Weber M, Rosenfeld W, et al. Heralded entanglement between widely separated atoms. Science. 2012; **337**(6090):72-75. Available from: http://science.sciencemag.org/content/337/6090/72

[41] Hensen B, Bernien H, Dréau AE, Reiserer A, Kalb N, Blok MS, et al. Loophole-free bell inequality violation using electron spins separated by 1.3 kilometres. Nature. 2015;**526**:682. DOI: 10.1038/nature15759

[42] Simon C, Irvine WTM. Robust long-distance entanglement and a loophole-free bell test with ions and photons. Physical Review Letters. 2003;**91**: 110405. DOI: 10.1103/PhysRevLett.91.110405

[43] Yin J, Cao Y, Li YH, Liao SK, Zhang L, Ren JG, et al. Satellite-based entanglement distribution over 1200 kilometers. Science. 2017;**356**(6343): 1140-1144. Available from: http://science.sciencemag.org/content/356/6343/1140

[44] Mishra A, Albash T, Lidar DA. Finite temperature quantum annealing solving exponentially small gap problem with non-monotonic success probability. Nature Communications. 2018;**9**:2917. DOI: 10.1038/s41467-018-05239-9

[45] Agarwal GS. Vacuum-field Rabi Splittings in microwave absorption by Rydberg atoms in a cavity. Physical Review Letters. 1984;**53**:1732-1734. DOI: 10.1103/PhysRevLett.53.1732

[46] Herrera F, Spano FC. Cavity-controlled chemistry in molecular ensembles. Physical Review Letters. 2016;**116**:238301. DOI: 10.1103/PhysRevLett.116.238301

[47] Schneider C, Porras D, Schaetz T. Experimental quantum simulations of many-body physics with trapped ions. Reports on Progress in Physics. 2012;**75**: 024401

[48] Dicke RH. Coherence in spontaneous radiation processes. Physics Review. 1954;**93**:99-110. DOI: 10.1103/PhysRev.93.99

[49] Hepp K, Lieb EH. On the superradiant phase transition for molecules in a quantized radiation field: The dicke maser model. Annals of Physics. 1973;**76**(2):360-404

[50] Gu X, Huai SN, Nori F, Yx L. Polariton states in circuit QED for electromagnetically induced

transparency. Physical Review A. 2016; **93**:063827. DOI: 10.1103/PhysRevA.93.063827

[51] Guerin W, Araújo MO, Kaiser R. Subradiance in a large cloud of cold atoms. Physical Review Letters. 2016; **116**:083601. DOI: 10.1103/PhysRevLett.116.083601

[52] Klinder J, Kebler H, Wolke M, Mathey L, Hemmerich A. Dynamical phase transition in the open Dicke model. National Academy of Sciences of the United States of America. 2015; **112**(11):3290-3295

[53] Will SA, Park JW, Yan ZZ, Loh H, Zwierlein MW. Coherent microwave control of ultracold $^{23}Na^{40}K$ molecules. Physical Review Letters. 2016;**116**: 225306. DOI: 10.1103/PhysRevLett.116.225306

[54] Baumann K, Guerlin C, Brennecke F, Esslinger T. Dicke quantum phase transition with a superfluid gas in an optical cavity. Nature. 2010;**464**: 1301-1306

[55] Gómez-Ruiz FJ, Acevedo OL, Rodríguez FJ, Quiroga L, Johnson NF. Pulsed generation of quantum coherences and non-classicality in light-matter systems. Frontiers in Physics. 2018;**6**:92. DOI: 10.3389/fphy.2018.00092

[56] Gómez-Ruiz FJ, Mendoza-Arenas JJ, Acevedo OL, Rodríguez FJ, Quiroga L, Johnson NF. Dynamics of entanglement and the Schmidt gap in a driven light–matter system. Journal of Physics B: Atomic, Molecular and Optical Physics. 2017;**51**(2):024001. DOI: 10.1088/1361-6455/aa9a92

[57] Gómez-Ruiz F, Acevedo O, Quiroga L, Rodríguez F, Johnson N. Quantum hysteresis in coupled light–matter systems. Entropy. 2016; **18**(9):319. DOI: 10.3390/e18090319

[58] Lee CF, Johnson NF. Spin-glasses in optical cavity. Europhysics Letters. 2008;**81**:37004. Available from: http://stacks.iop.org/0295-5075/81/i=3/a=37004

[59] Jarrett TC, Lee CF, Johnson NF. Optically controlled spin-glasses in multi-qubit cavity systems. Physical Review B: Rapid Communications. 2006; **74**:121301. DOI: 10.1103/PhysRevB.74.121301

[60] Lee CF, Johnson NF. First-order super-radiant phase transitions in a multi-qubit–cavity system. Physical Review Letters. 2004;**93**:083001. DOI: 10.1103/PhysRevLett.93.083001

[61] Jarrett TC, Olaya-Castro A, Johnson NF. Optical signatures of quantum phase transitions in a light-matter system. Europhysics Letters. 2007;**77**:34001. Available from: http://stacks.iop.org/0295-5075/77/i=3/a=34001

[62] MacKay N. When Lanchester met Richardson, the outcome was stalemate: A parable for mathematical models of insurgency. Journal of the Operational Research Society. 2015;**66**:2

[63] Fratini M et al. Scale-free structural organization of oxygen interstitials in LaCuO. Nature. 2010;**466**:841

[64] Johnson NF et al. Equivalent dynamical complexity in a many-body quantum and collective human system. AIP Advances. 2011;**1**:012114

[65] Nagy D, Kónya G, Szirmai G, Domokos P. Dicke-model phase transition in the quantum motion of a Bose-einstein condensate in an optical

cavity. Physical Review Letters. 2010; **104**:130401. DOI: 10.1103/PhysRevLett.104.130401

[66] Das P, Tasgin ME, Müstecaplõoğlu OE. Collectively induced many-vortices topology via rotatory Dicke quantum phase transition. New Journal of Physics. 2016; **18**:093022. DOI: 10.1088/1367-2630/18/9/093022

[67] Hioe FT. Phase transitions in some generalized Dicke models of Superradiance. Physical Review A. 1973; **8**:1440-1445. DOI: 10.1103/PhysRevA.8.1440

[68] Emary C, Brandes T. Chaos and the quantum phase transition in the Dicke model. Physical Review E. 2003;**67**: 066203. DOI: 10.1103/PhysRevE.67.066203

[69] Betancourt JM, Rodríguez FJ, Quiroga L, Johnson NF. Ladder of Loschmidt anomalies in the deep strong-coupling regime of a qubit-oscillator system. Physical Review A. 2021;**104**: 043712. DOI: 10.1103/PhysRevA.104.043712

[70] Méndez-Córdoba FPM, Rodríguez FJ, Tejedor C, Quiroga L. From edge to bulk: Cavity-induced displacement of topological nonlocal qubits. Physical Review B. 2023;**107**: 125104. DOI: 10.1103/PhysRevB.107.125104

[71] Acevedo OL, Quiroga L, Rodríguez FJ, Johnson NF. Large dynamic light-matter entanglement from driving neither too fast nor too slow. Physical Review A. 2015;**92**: 032330. DOI: 10.1103/PhysRevA.92.032330

Section 3

Data Representation and Quantum Information Processing

Chapter 6

Unitary Maps and Quantum Artificial Neural Networks

Carlos Pedro Gonçalves

Abstract

Unitary quantum maps provide a bridge between classical and quantum dynamical systems theories, having been applied within the context of quantum chaos research. When applied to quantum artificial neural networks, as models of networked quantum computation, unitary quantum maps allow one to address these networks as quantum networked dynamical systems. In this chapter, we address the application of these maps to quantum artificial neural networks, specifically studying the simulation and implementation of these maps for quantum recurrent neural networks, simulating these networks as dynamical computational systems and researching the topological properties of the series of neural firing operators' quantum averages for nonstationary network states. We also research the results of a simulation of one of these networks on a quantum computer by cloud-based access to IBM Q Experience resources. The results show the emergence of complex dynamics, fitting into similar classes as those of classical cellular automata and coupled maps, including topological markers of chaos, edge of chaos and fractal attractors in the sequences of quantum averages. The implications for quantum complexity research, quantum chaos theory and quantum computing are addressed.

Keywords: quantum artificial neural networks, quantum complexity, unitary quantum maps, quantum computation, quantum dynamical systems, quantum chaos

1. Introduction

Complexity research deals with complexity in different contexts including physical systems, biological systems, social, economic, financial and even technological systems. Despite this diversity of contexts, there is a core disciplinary basis in complexity research that intersects computer science, dynamical systems science, statistical mechanics and network science [1–7]. The dialog between these research fields that form the core disciplinary basis of complexity research has led to fundamental research developed around general families of models of computing systems based on networked computation, these include: cellular automata, coupled maps, random Boolean networks and neural networks [2, 5–7].

Several of these models have their quantum counterparts and form an important research direction within quantum complexity research, associated with quantum networked computation and artificial intelligence (AI) [8–12].

Early work in quantum complexity has included the contributions of Ilya Prigogine and Hermann Haken's groups, focusing on far-from-equilibrium dynamics, irreversibility, dissipative structures and synergetics [3, 4].

In the case of Ilya Prigogine's group, the relevance of quantum complexity research led to the transformation of *The Ilya Prigogine Center for Studies in Statistical Mechanics and Complex Systems* into *The Center for Complex Quantum Systems*, which includes research into quantum chaos [3, 13, 14].

Other quantum complexity research groups have since developed, an example being the Richter Group on Complex Quantum Systems, from the University of Regensburg, which has also worked on many bodies' quantum systems, quantum chaos theory and, also, control of quantum chaos [15–17].

Of the different research lines in quantum complexity, the research on the dynamics of quantum artificial neural networks, or quantum neural networks (QNNs) for short, and on quantum cellular automata (QCAs) leads to a direct bridge between fundamental research within complexity science and quantum computer science, with implications for quantum technologies, linked to quantum networked computation with relevance for distributed AI systems and automation in quantum networks.

The current chapter addresses QNNs. The QNNs are models of networked quantum computation and have been studied by different authors [11, 12, 18–22]. They can be employed both in the context of quantum machine learning [18–22] and as models of quantum networked dynamical systems [12, 22], in this case, for artificial neurons with two-level neural firing patterns, a network with n-neurons corresponds to an n-qubit quantum computing system. In this context, neural firing patterns can be represented by neural firing operators, which correspond to the projectors onto the basic binary computational basis that spans an n-qubit system.

The difference with respect to a general n-qubit quantum computing system is that the network's unitary computation is comprised of conditional unitary operators, that is, different local unitary operations are applied for different neural firing patterns, in this way, the sequence of unitary transformations associated with a QNN, can be placed in the form of general quantum circuits in terms of representation of the conditional unitary operators, which establishes a bridge to standard quantum computation [20, 21].

While these works address specifically QNNs as running quantum algorithms that end at a specific step, another type of application involves a bridge to quantum dynamical systems, leading to the need to work with the concept of a quantum map.

Indeed, a feature of classical artificial neural networks is that they can be used to simulate complex networked dynamical systems, the quantum extension of this type of application, involves performing the conditional unitary operations of the network running the full activation cycle, leading to a new state of the network at the end of each computational cycle, as addressed in [12, 22].

In this way, each computational cycle corresponds to an iteration of the network and a sequence of iterations of the network can be represented by the concept of unitary quantum map [12].

The concept of unitary quantum map was introduced in quantum chaos research to deal with systems like the kicked top and has become a standard research tool of quantum chaos research [23].

This concept is an extension of the concept of a classical map, from classical dynamics, that maps a classical phase space onto itself and can be used to express the classical transformation of a state into another state leading to a discrete-time sequence of iterations [23].

A unitary quantum map is, in turn, a unitary operator defined on a quantum system's Hilbert space that maps a state vector into another state vector, leading to a sequence of state vectors, this sequence can be researched, using the quantum theory's formalism, by calculating the quantum averages of observables and researching this sequence of quantum averages. A major feature of a unitary quantum map is that, being unitary, it maps pure states into pure states.

In the case of QNNs, when addressed as basic quantum dynamical systems, the concept of unitary quantum map becomes relevant, since we are not running the network a single time with a measurement at the end of a computational cycle, instead one runs the sequence of iterations of the quantum map that corresponds to a full cycle of computations of the network [12].

It turns out that, using this methodology, specific complex dynamics have been identified in these networks, including complex emergent attractors in the sequences of quantum averages linked to neural firing patterns as well as topological features of chaos-like dynamics [12, 22].

In order to study the network's dynamics, one can use basic quantum formalism, like in quantum chaos theory, and employ matrix calculus to calculate, at the end of each iteration of the map, the quantum averages for relevant observables on the network's activity such as neural firing patterns.

In the case of neural firing patterns, since they are represented by the basic projectors onto the network's computational basis, the sequence of quantum averages has a specific interpretation, namely, the quantum averages of these projectors onto the computational basis, at the end of t iterations of the network provide the probability distribution associated with the network's alternative firing pattern if a measurement was performed at the end of those t iterations.

This last property allows, as we show in the present chapter, for an empirical testing of the theoretical sequence of quantum averages obtained from the sequence of unitary state vectors using the formalism of quantum mechanics.

By running the network for a sequence of iterations, which is a unitary stage, and then extracting frequencies obtained by measuring each quantum register's state at the end, the results, when run for multiple repeated experiments should match the theoretical quantum averages. This can be done for different iteration indexes t which allows for an extraction of the sequence of quantum averages which can be compared with the theoretical sequence obtained from matrix calculus, using the basic formalism of quantum mechanics.

In order to characterize the network's dynamics one can apply topological analysis methods, namely, recurrence analysis to the sequences of quantum averages. This approach was followed in [12, 22] to characterize these networks' dynamics, both in the case of a dissipative quantum map [22] and in the case of a unitary quantum map [12].

When considering recurrence analysis, there are four main classes of topological signatures, as we will review in the present chapter, that mirror four main dynamical classes identified in cellular automata [6], and that also occur in coupled map lattices [5] and in random Boolean networks [7], it turns out that these four dynamical classes are shown to also occur in QNNs [12, 22]. These classes are:

- *Class 1*: The fixed point, which is a stationary state, corresponds to the eigenvectors of the quantum map for the network, which usually involves entangled states [12];

- *Class 2*: The periodic or quasiperiodic dynamics;
- *Class 3*: The random dynamics, which can occur in deterministic systems including in cellular automata and in deterministic chaos [5, 6], and that also occurs in QNNs as we will see in the present chapter, especially when the number of neurons is increased;
- *Class 4*: Finally, there is the fourth class of dynamics which includes a mixture of class 2 and class 3, and that has been called the *edge of chaos* [6, 7, 24, 25].

Class 4 dynamics can occur in chaotic dynamical systems near the onset of chaos, in cellular automata, in random Boolean networks, as well as in coupled map lattices, as propagating localized order that intermixes with random fluctuations characterizing phenomena such as spatiotemporal intermittency [5–7, 24, 25], and also occurs in the sequences of quantum averages of neural activity operators for QNNs being identifiable through the application of recurrence analysis methods as researched in detail in [12, 22, 26], showing the effectiveness of topological analysis methods [27, 28], which have now been expanded and adapted to quantum chaos research [12, 22, 26, 29].

In the current chapter, we illustrate these dynamics for different QNNs, focusing on class 4 (edge of chaos) and class 3 (random) dynamics.

In Section 2, we review the concept of unitary quantum map and its application to QNNs, this section is divided into multiple subsections.

In Section 2.1, we address the concept of unitary quantum map and how it can be studied both theoretically and experimentally in general.

In Section 2.2, we address the example of a unitary quantum map's dynamics for a single qubit and illustrate the basic application of recurrence analysis methods on the theoretical simulation of the map.

In Section 2.3, we review a two-neuron recurrent neural network that has been researched in [12, 26] and in Section 2.4, we address the comparison between the theoretical simulation of the model and the empirical implementation on one of IBM's quantum computers, accessed by cloud using IBM Q Experience resources.

In Section 3, we address the dynamics of a QNN with a ring topology which includes elements of a feedforward network with a recurrent connection at the end, also providing for a bridge to quantum cellular automata and quantum deep neural networks. We simulate this network for different numbers of neurons showing the transition from class 4 to class 3.

In the case of the ring topology network and for the IBM quantum computer simulation, we also uncover the emergence of fractal attractors, a point that is also discussed in Section 3.

In Section 4, we conclude with a final reflection on the chapter's findings and its implications for quantum complexity research, quantum technologies and quantum chaos theory.

2. Quantum artificial neural networks as dynamical quantum computing systems: theory and experiments

2.1 Unitary quantum maps and quantum dynamics

In classical dynamical systems, a map is a function that maps the phase space onto itself, leading to a discrete-time sequence of states. In quantum theory's formalism, a quantum extension of a classical map is a quantum unitary map, defined on a quantum system's Hilbert space $\mathbb{H}$, that maps quantum state vectors into quantum state vectors [23]:

$$|\psi(t+1)> \; = F|\psi(t)> \tag{1}$$

In this context, $t = 0, 1, 2\ldots$ is an iteration index, leading to a sequence of quantum states $|\psi(t)>$.

To research quantum dynamics with respect to dynamical classes, the sequences of states themselves are not the most useful since the concept of trajectory breaks down, however, when an observable is used, like a position, momentum, Hamiltonian operator and spin operator, one can effectively study the sequence of quantum averages for that observable by employing basic quantum matrix calculus, so that, given the sequence of quantum states one can extract a form of trajectory for the observable:

$$\langle O \rangle_t = \langle \psi(t)|O|\psi(t)\rangle \tag{2}$$

This is a well-known basic scheme used in both the Schrödinger and Heisenberg picture in quantum complexity research [30]. However, it should be stressed that, in this scheme, which has been extensively applied to the study of quantum dynamics [30], one is not measuring the observable, instead one is using quantum matrix calculus to extract a sequence of quantum averages that provides a picture of a quantum trajectory in terms of an observable.

Of course, the sequence of quantum averages provides an important information in what regards the empirical implications, namely, $\langle O \rangle_t$ corresponds to the theoretical prediction of the expected value of the observable if the iterations were to be stopped at step t and the observable was measured at the end of that step.

Indeed, if one were to perform the measurement on multiple identical copies of the system, the sample average should approximate the quantum average for a large number of experiments, this is a basic point that comes out of the early empirical testing of quantum theory.

Of course, if a measurement is performed at the t-th iteration, the unitarity will break down at that point, so we are effectively ending the iteration sequence at step t. This point has an important implication for the theoretical versus empirical research on the dynamics resulting from quantum unitary maps.

In the theoretical research, the sequence of quantum averages for an observable resulting from the sequence of quantum states can be calculated and dealt with as a form of trajectory using the basic formalism of quantum mechanics, as stated above, providing another picture of the unitary evolution itself in regards to the system in question and a relevant observable, a point that is extensively addressed in the quantum noise research [30].

The theoretical value of the quantum average at iteration step t provides, as stated above, for a reference on what the quantum average would be for the observable if we performed a quantum measurement of that observable, for a large number of identically prepared systems, at that iteration step, ending the iterations at that step.

In this way, while the theoretical simulation of the map can either be extracted as a closed formula or numerically using standard matrix calculus, in order to obtain the physical empirical testing of the quantum map's iterations, one must use an experimental loop comprised of the following steps that we explain next:

1. For $t = 1, 2, \ldots, T$:

   i. For $n = 1, 2, \ldots, N$:

      a. Set the initial quantum state for: $|\psi(0)>$.

      b. Perform the unitary evolution defined by applying the map s times ($|\psi(s)> = F^t|\psi(0)>$)

      c. Perform a quantum measurement of the observable at the end of iteration t and store the result.

   ii. Calculate the average of the results stored at the end of each step 1.1.2.2. and keep it.

   iii. Return to step 1.1.

The above general experimental setup means that, for each sequence of iterations t, we perform the unitary evolution t times on an ensemble of N identical copies of the system in question for the same initial condition and then measure the observable at the end of the iteration step t, calculating the empirical average for that ensemble for that iteration step, in this way we get a sequence of quantum averages that will closely match the theoretical sequence, given quantum theory's predictions. This is a basic extension of the standard empirical testing of quantum mechanics applied to the general context involving unitary maps.

Now, considering the specific case of an observable, with a discrete eigenvalue spectrum, such that its eigenvectors span the Hilbert space $\mathbb{H}$:

$$O|v_k> = v_k|v_k> \tag{3}$$

We can introduce the complete set of projectors onto the observable's basis:

$$P_O = \{P_k = |v_k> <v_k| : k = 0, 1, \ldots\} \tag{4}$$

In this case, given the sequence of state vectors $|\psi(t)>$, employing the formalism of quantum mechanics [30], we can calculate the sequence of quantum averages for the observable and research the corresponding dynamical sequence:

$$\langle O \rangle_t = \langle \psi(t)|O|\psi(t)\rangle \tag{5}$$

We can also study the sequence of quantum averages for each projector:

$$\langle P_k \rangle_t = \langle \psi(t)|P_k|\psi(t)\rangle \tag{6}$$

In the case of a finite-dimensional basis, which leads to a finite-dimensional projector set and corresponding Hilbert space, we can embed the quantum averages of the projectors into a d-dimensional Euclidean space:

$$P(t) = (\langle \psi(t)|P_0|\psi(t)\rangle, \langle \psi(t)|P_1|\psi(t)\rangle, \ldots, \langle \psi(t)|P_{d-1}|\psi(t)\rangle) \tag{7}$$

The above tuple is a probability tuple containing the probability distribution associated with each alternative eigenstate, calculated for the t-th iteration step, in this way, the sum of the elements in the tuple is equal to 1. Since the sequence of values for $P(t)$ is a d-dimensional tuple of real numbers, the sequence can be represented and studied in a d-dimensional Euclidean space.

Again, as for an observable's quantum average, the above tuple's values can be extracted by applying basic matrix calculus employing quantum mechanics' formalism and can be simulated in any software that has linear algebra features, including Python's NumPy, MATLAB or Mathematica.

The theoretical simulation of the tuple sequence provides for a sequence of values of the quantum probability distribution, again, as in the case of the quantum average of an observable, the value of the tuple at iteration step t has an interpretation as the probability distribution associated with interrupting the iterations at that iteration step and performing a quantum measurement of the observable, at that step.

For repeated identically prepared experiments, one can extract the statistical frequencies for the different eigenstates at the end of the t iterations, stopping the iterations at that point, so that the frequency distribution for a large number of experiments will match the theoretical results, this is a basic approach that has been followed since the beginning of quantum mechanics and that carries over to the experimental implementation of quantum dynamics of quantum maps.

In order to study the dynamics of the quantum probability tuple $P(t)$ or of an observable's sequence of quantum averages, one can apply topological data analysis methods, in particular, one can apply recurrence analysis methods, that are effective in distinguishing between dynamical regimes [27, 28].

The analysis is based on calculating a distance matrix between each dynamical point in Euclidean space, this matrix is symmetric and the main diagonal is comprised only of zeros. From the distance matrix, one can estimate a binary recurrence matrix that stores the value 1 when two dynamical points in a sequence are distant from each other by at most a radius of ε and 0 when two points are distant from each other by more than ε.

The recurrence matrix can also be worked with as a transition matrix, such that each dynamical point is linked to all the neighboring points in a closed Euclidean ε-neighborhood.

The use of the closed neighborhood, instead of the open neighborhood, allows the distinction between periodic and nonperiodic orbits, in the case of a periodic orbit, a closed neighborhood is non-empty for a radius of zero between periodic points, while for nonperiodic orbits, if the radius is set to zero, the recurrence matrix will be a null matrix.

Periodic dynamics appear when the radius is set to zero, as evenly spaced diagonals parallel to the main diagonal that are completely filled with the value of 1. This is no longer the case for nonperiodic dynamics, in this last case, when the dynamics is quasiperiodic, long unevenly spaced diagonals filled with the value of 1 appear for a sufficiently high radius.

Topological markers of chaos appear as interrupted diagonals and isolated dots, these interrupted unevenly spaced diagonals result from the exponential divergence associated with positive Lyapunov exponents.

When considering the types of dynamics, there are four main recurrence classes that occur in dynamical systems that closely mirror cellular automata's four classes reviewed in the introduction [6, 12, 22], these are:

- *Class 1* (*fixed point*): the recurrence matrix is a unit matrix, all points are recurrence points.
- *Class 2* (*periodic or quasiperiodic dynamics*): the dynamics is either periodic with evenly spaced diagonals matching the period or quasiperiodic, which shows up as unevenly spaced diagonals.
- *Class 3* (*random*): the random dynamics can either result from external noise or be endogenously generated in a deterministic system, as it happens in the case of deterministic chaos and class 3 cellular automata, in these cases, the dynamics is actually pseudorandom and its topological markers can include broken diagonals, linked to an unstable periodic orbit skeleton and disconnected points in the recurrence matrix.
- *Class 4* (*edge of chaos*): this is a mixture of class 2 and class 3, its major characteristic in black and white recurrence plots include very long resilient unbroken diagonal lines that appear with different distances along with broken diagonals and isolated dots that represent random-like signatures, QNNs tend to produce this dynamical class.

In the case of chaotic dynamics for shift maps, the randomness is associated with the randomness of the binary expansion of the initial condition, which differs from class 3 random cellular automata that can produce random patterns even for non-complex initial conditions, this is a point stressed in [6].

In networks of chaotic oscillators, such as coupled map lattices, class 3 and class 4 behavior also occur in these networks' dynamics, with class 4 type dynamics occurring, for instance, as spatiotemporal intermittence [5].

It is important to stress that there is a continuum between class 3 and class 4, in the case of classical chaotic dynamical systems, in the sense that, in the chaotic phase, but near the onset of chaos, with lower values of positive Lyapunov exponents, pockets of unstable cycles can be mirrored for a longer time and lead to class 4-type topological markers, these markers, as stated, also occur in networks of chaotic oscillators [5].

In this way, the concept of edge of chaos that was addressed in both the analysis of cellular automata and random Boolean networks [6, 7] can also apply to deterministic chaos near the onset of chaos and networks of chaotic oscillators [5].

Now, in order to illustrate the application of the concept of unitary quantum map, in the context of quantum computation, we begin with a single qubit example and a Walsh-Hadamard transform.

2.2 Single qubit dynamics: the example of the Walsh-Hadamard transform

Returning to the quantum dynamics, as an example of a quantum map applied to a single qubit computational system, let us consider the Walsh-Hadamard transform:

$$|\psi(t+1)> \; = W|\psi(t)> \tag{8}$$

$$W = \frac{|0><0| - |1><1| + |0><1| + |1><0|}{\sqrt{2}} \tag{9}$$

Starting with $|\psi(0)> = |0>$, and working with the projectors $P_0 = |0><0|$ and $P_1 = |1><1|$, the sequence $P(t)$, for the above initial condition, is periodic cycling through the tuples $(0.5, 0.5)$ and $(1, 0)$, with the first tuple holding for even iterations and the second for the odd iterations. This means that if we stop the iterations at an odd t and were to measure the qubit at that step, then, one would obtain the logical state 0, with probability equal to 1, on the other hand, if were one to stop the iterations at an even t and were to measure the qubit at that step then one would obtain the logical state 0, with probability 0.5, and the logical state 1, with probability 0.5.

In **Figure 1** (right), we plot the results of the iterations of the above quantum map as a scatterplot for the two projectors' quantum averages with respect to the iteration index, which shows the cycle through the above values, also in **Figure 1** (left) we show the corresponding recurrence plot which plots the recurrence matrix for a radius of zero, the plot visibly shows the periodic orbit obtained for the quantum averages with the long fully filled diagonals surfacing for this radius.

The result is a regular grid-like pattern for the recurrence plot. Since the matrix is symmetric the full recurrence structure can be researched using either the diagonals above the main diagonal or below, for which different metrics can be calculated.

The first metric that we use is to count the number of diagonals above the main diagonal with recurrence points, this number corresponds to the total number of diagonal lines above the main diagonal with recurrence points.

The second counting is the number of diagonals above the main diagonal with 100% recurrence, that is, a diagonal that is fully filled with recurrence points, finally, the last counting is the number of diagonals above the main diagonal.

From these three metrics, one can calculate the recurrence probability, that is, the probability that a randomly chosen diagonal above the main diagonal contains recurrence points, this metric can be calculated by dividing the number of lines above the main diagonal with recurrence points by the total number of diagonals above the main diagonal.

The second metric that can be calculated is the probability that a randomly chosen diagonal above the main diagonal with recurrence points is a 100% recurrence diagonal, this conditional probability can be calculated by dividing the number of diagonals above the main diagonal with 100% recurrence by the number of diagonals above the main diagonal with recurrence.

A third metric that can be calculated is the average recurrence strength. In this case, one starts by dividing the number of recurrence points in a diagonal with recurrence by the length of the diagonal, which provides for the proportion of the

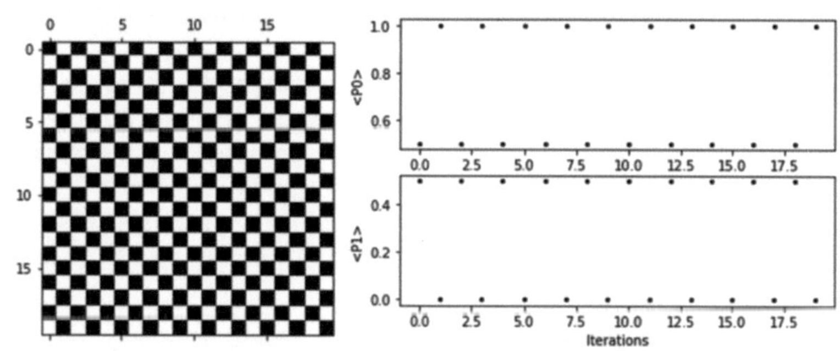

Figure 1.
Recurrence plot (left) for a radius of 0 from a 20 iterations' simulation of the quantum Walsh-Hadamard map for a single qubit, scatterplots (right) for the sequence of quantum averages of the two computational basis projectors.

Metrics	
Number of lines with 100% recurrence	9
Number of lines with recurrence	9
Number of diagonals	19
Average recurrence strength	1
P[Recurrence]	0.4737
P[100%recurrence|recurrence]	1

Table 1.
*Recurrence metrics for the recurrence plot shown in **Figure 1**.*

diagonal that is filled with recurrence points, this proportion is 0 for a diagonal with no recurrence points and 1 for a 100% recurrence diagonal.

After calculating this proportion for each diagonal above the main diagonal, one can calculate the mean of this proportion which provides for the average recurrence strength.

Finally, we research the distribution of distances between diagonals with 100% recurrence, measured in terms of the number of iterations needed to reach the next 100% recurrence diagonal, this leads to the periodicity of a cycle associated with 100% recurrence diagonals and becomes a critical statistic for addressing possible cycles, or, in the case of nonperiodic dynamics, the possibility of quasiperiodic structures.

This distribution will play a major role in the analysis of the class 4 dynamics as we will see.

It turns out that these metrics are effective as we will see in distinguishing between the four dynamical classes.

If we calculate these metrics for **Figure 1**'s simulation, we obtain the results shown in **Table 1**.

The recurrence metrics reflect the periodic structure, in this case, there were 20 iterations. The distance matrix is, therefore, a square matrix of rank 20, which leads to 19 diagonals above the main diagonal, the cycle is period 2, which means that we get 9 diagonals all with full recurrence, the probability of a randomly chosen diagonal with recurrence being a 100% recurrence diagonal is, therefore, equal to 1, on the other hand, the probability of a diagonal being a diagonal with recurrence points is equal to 9/19 which is approximately 0.4737.

The average recurrence strength is equal to 1 since all diagonals with recurrence are 100% recurrence diagonals.

Now, since there are nine 100% recurrence diagonals and it takes two iterations to reach the next 100% recurrence diagonal, the distances between 100% recurrence diagonals are all equal to 2 and occur eight times, as expected, from a period 2 dynamics. The above dynamics is an example of a class 2 dynamics.

A more complex dynamics is obtained when dealing with QNNs, as an example, we now address a two-neuron quantum recurrent neural network.

2.3 Two-neuron recurrent neural network

Considering a two-neuron quantum recurrent neural network with two-level firing patterns, the general architecture is shown in **Figure 2**.

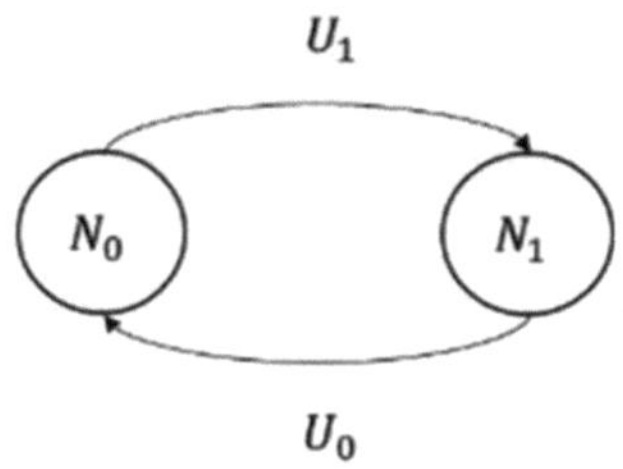

Figure 2.
Quantum recurrent neural network architecture.

In this case, the Hilbert space for the network is spanned by the computational basis $B_2 = \{|00>, |01>, |10>, |11>\}$, such that $|rs>$ with $r, s \in \{0, 1\}$ encodes a neural firing pattern, that is, if r is equal to 0 then the first neuron is not firing, while, if it is equal to 1, the first neuron is firing, the same holds for s, in relation to the second neuron.

The main projectors for this basis lead to the neural firing pattern projectors $P_{rs} = |rs> <rs|$, which means that the quantum averages' tuple $P(t)$ can be embedded in four-dimensional Euclidean space. For each neuron, we can also calculate the local neural firing operators $N_0 = P_1 \otimes I$ and $N_1 = I \otimes P_1$, where I is the rank 2 identity matrix, the first operator yields the value of 1 when the first neuron is firing and 0 otherwise, and the second operator yields the value of 1 when the second neuron is firing and 0 otherwise.

In the network that we will be addressing, the two operators U_1 and U_0 are conditional operators that conditionally transform the quantum register of the target neuron, conditional on the firing pattern of the input neuron, following the network's links:

$$U_0 = I \otimes |0> <0| + U_r \otimes |1> <1| \tag{10}$$

$$U_1 = |0> <0| \otimes I + |1> <1| \otimes U_r \tag{11}$$

$$U_r = \cos\left(\frac{\pi r}{2}\right)(|0> <0| + |1> <1|) + \sin\left(\frac{\pi r}{2}\right)(-|0> <1| + |1> <0|) \tag{12}$$

Assuming that the first neural connection that is activated is from the first neuron to the second and that the second neural connection to be activated is from the second neuron to the first, running the network through a full computational cycle is given by the operator product:

$$F = U_0 U_1 \tag{13}$$

Repeated sequential full computational cycle operations for this network lead to the quantum map:

$$|\psi(t+1)> = F|\psi(t)> = U_0 U_1 |\psi(t)> \tag{14}$$

To illustrate the dynamics we will work here with the initial superposition state:

$$|\psi(0)> = \frac{|00> + |01> + |10> + |11>}{2} \tag{15}$$

The dynamics of this network is already nontrivial, it depends upon the value of r. For low values of r, the quantum averages for the projectors in which at least one neuron is active track down a sinusoidal curve.

In **Figure 3**, we show the corresponding sequence for the quantum averages associated with the projector set. It should be stressed that the figure is actually a scatterplot, for a discrete-time sequence.

From the results shown in **Figure 3**, a few points can be noticed, first, the quantum average for the projector where the two neurons are nonfiring is constant as can be seen in the top graph, this is a class 1 dynamics.

The quantum averages for the remaining projectors follow the periodic sinusoidal curves which are class 2 dynamics, in this case, when the value rises for the projector where the two neurons are active, which corresponds to a reinforcing neural firing activity, it falls for the projectors where the two neurons are in reverse neural firing activity, which corresponds to inhibitory neural firing activity, on the other hand, the quantum averages, for the cases where the two neurons show reverse neural activity, follow a synchronized sinusoidal curve.

As shown in [12, 26], for the values of r that lead to class 2 dynamics, the dynamics for the averages of the local neural firing operators N_0 and N_1 are synchronized and also follow a sinusoidal curve.

As the parameter r is increased the period associated with the sinusoidal curves decreases as well as the correlation between the quantum averages for the two-neurons' local neural firing operators, until there is a transition from positive to negative correlation. The point at which the dynamics transitions from positive to negative correlation is also a point after which the dynamics no longer follows the sinusoidal periodic curve transitioning to dynamical regimes characterized by class 4 dynamics [12, 26].

While the quantum average for the projector P_{00} remains in its fixed point dynamics for all values of r, for the remaining projectors, two major attractor shapes emerge with the increase of the parameter r and the disappearance of the sinusoidal pattern.

The main attractor has a triketa-like shape, which can be obtained from intersections of three ellipsoids, and is largely characterized by class 4 dynamics, the other has

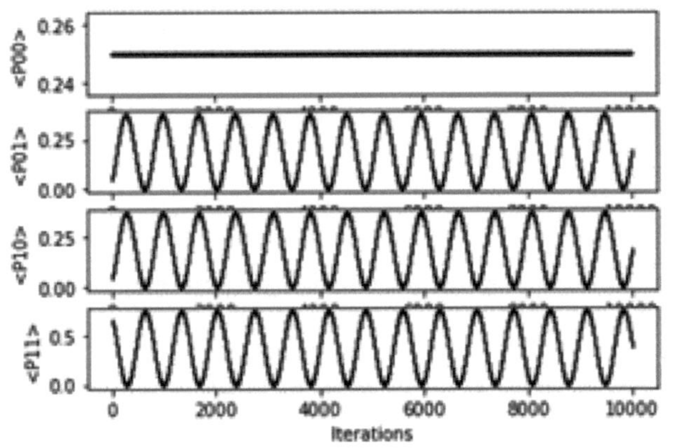

Figure 3.
Simulation of the quantum recurrent neural network with r *equal to 0.002, 10,000 iterations are shown after the first 1000 iterations were dropped for transients, the trajectory of each component of the projectors' tuple* P(t) *is shown.*

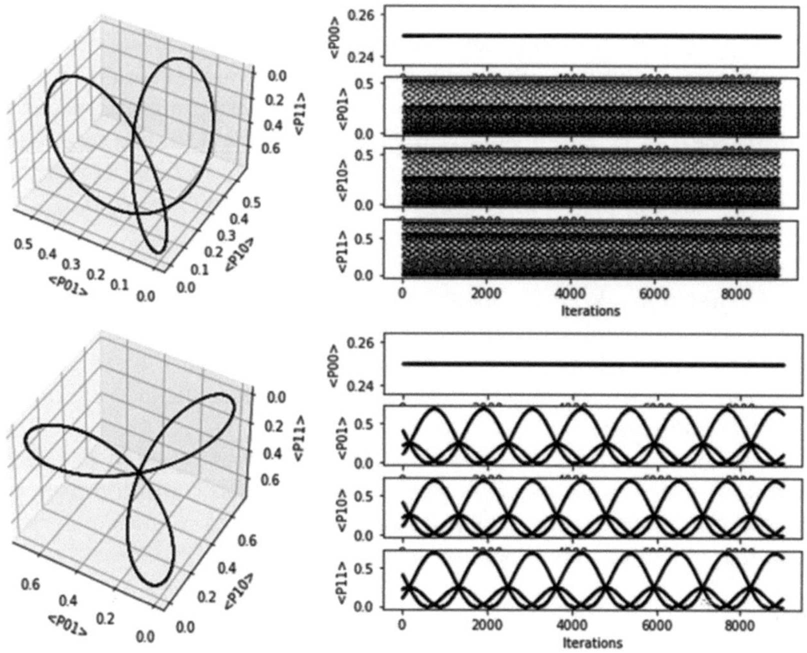

Figure 4.
Simulation of the quantum recurrent neural network with r = 0.550129597 *(top) and with* r = 0.999 *(bottom), 10,000 iterations are shown in each case after the first 1000 iterations were dropped for transients, the scatterplot trajectory of each component of the projectors' tuple* P(t) *is shown (right) along with the attractor (left).*

the shape of a trifolium, the trifolium is characterized by class 2 dynamics resulting from a progressive change in the triketa attractor which loses its middle section leading to the trifolium as r is increased.

In **Figure 4**, we show both attractor shapes and corresponding quantum averages for the projectors shown in scatterplots, in the case of the triketa, the plotted dynamics is for a parameter with near-zero correlation between the sequence of quantum averages of N_0 and N_1.

2.4 Experimental implementation of the two-neuron recurrent neural network

The experiments in the quantum device involve the standard application of quantum measurement for a quantum map, which is discussed in Section 2.1. In this case, for each value of t, in order to extract the sequence $P(t)$, we need to apply the quantum map t times, and then measure the two quantum registers' states at the end of t, this needs to be repeated multiple times (multiple rounds) in order to obtain the empirical frequency distribution from which $P(t)$ can be estimated calculating the relative frequencies for the different computational basis values.

For each iteration index t, the t applications of the quantum map are automatically converted into a quantum circuit that implements the transformation F^t, this process is automated in IBM's systems, and the Qiskit code for the Jupyter Notebook used for the experiments NSimul.ipynb is provided on Github at the QNeural project's website [31, 32], the project which also contains the simulation basis for the theoretical model and that was used in [12, 22, 26].

Figure 5 shows 1000 iterations of the quantum map performed on IBM's quantum device "ibm_q_belem," for the triketa attractor.

As can be seen in **Figure 5**, the resulting sequence of values for $P(t)$ matches well the theoretical sequence, however, there are random fluctuations associated with both the quantum measurement and environmental noise, which, in this case, while low, can still lead to fluctuations in the relative frequencies for each measurement, this is evident in the cloud of points indicating a dispersion, we will return to this point further on.

In **Figure 6** (left), we show the colored recurrence plot for the triketa attract0r's theoretical simulation, also using 1000 iterations, and in **Figure 6** (right), we show the corresponding colored recurrence plot for **Figure 5**'s simulation.

The two recurrence plots are very similar and both exhibit the presence of a quadratic grid disturbed by random fluctuations, the quadratic grid is evidence of the presence of periodicities.

The parameter for which the simulation is produced, as stated, is one in which the local neural firing operators' sequence of quantum averages transition from positive to negative correlation, with their dynamics embedded in two-dimensional space corresponding to a two-dimensional projection of the above attractor [12, 26].

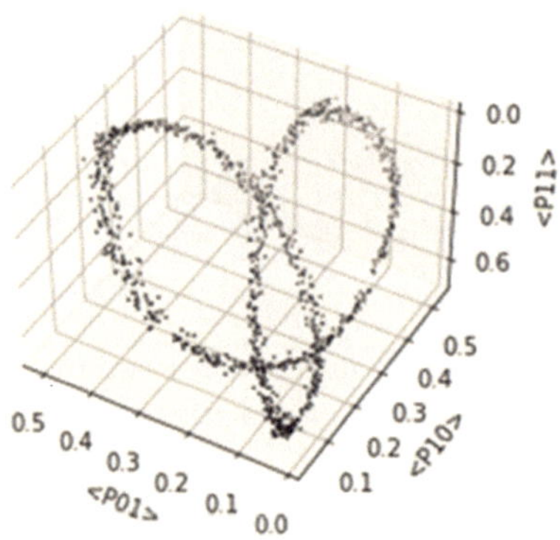

Figure 5.
Simulation of the quantum recurrent neural network on the "ibm_q_belem" device, for r = 0.550129597, *with* 5000 *rounds sent for each iteration and* 1000 *iterations kept after the first* 1000 *iterations were dropped for transients.*

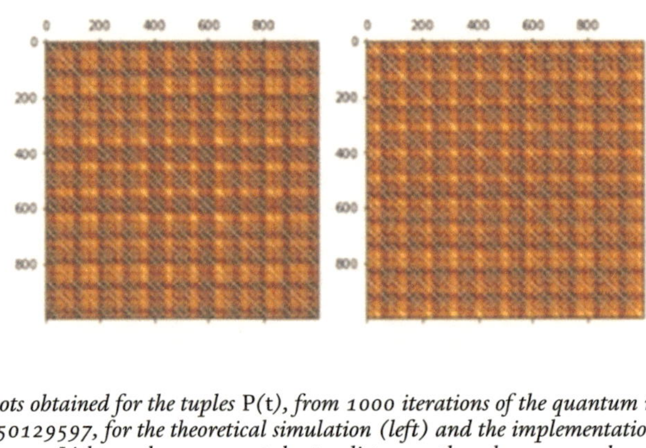

Figure 6.
Colored recurrence plots obtained for the tuples P(t), *from* 1000 *iterations of the quantum recurrent neural network with* r = 0.550129597, *for the theoretical simulation (left) and the implementation on "ibm_q_belem" (right) shown in* **Figure** 5. *Lighter colors represent shorter distances, the color map used was Python's matplotlib's cm.inferno_r.*

Metrics	Theoretical	Empirical
Number of lines with 100% recurrence	156	135
Number of lines with recurrence	964	978
Number of diagonals	999	999
Average recurrence strength	0.9558	0.9289
P[Recurrence]	0.9650	0.9790
P[100%recurrence|recurrence]	0.1618	0.1380

Table 2.
*Recurrence metrics for the recurrence matrices of **Figure 6**'s theoretical simulation and empirical simulation obtained for a Euclidean radius of 0.1.*

In **Table 2**, we provide the main recurrence metrics for both the theoretical simulation and the empirical results from the quantum device, using a radius of 0.1. In this case, for 1000 iterations, the differences between the theoretical simulation and the physical device's implementation are not large, the values are close to each other.

The main differences are that the physical device simulation has a slightly higher number of lines with recurrence, but it has a lower number of lines with 100% recurrence, leading to a higher recurrence probability but a lower probability of a randomly chosen line with recurrence being a line with 100% recurrence, the average recurrence strength is also smaller for the physical device.

This is consistent with a higher number of interrupted diagonals in the physical device implementation which is, in turn, consistent with the presence of noise associated with the physical device and the quantum measurement itself.

In both cases, the probability of a randomly chosen diagonal to contain recurrence points is high, however, the probability of a randomly chosen diagonal with recurrence points to be a 100% recurrence line is low, which indicates that the majority of lines with recurrence are either broken lines or contain disconnected dots, the dominance, in this case is of interrupted diagonals (unstable cycles), since the average recurrence strength in both the theoretical simulation and the empirical implementation in the physical device have a higher than 90% average recurrence strength, which indicates that, on average, the diagonals with recurrence have more than 90% of their size occupied by recurrence points, for this radius. This indicates that there are strong recurrence signatures.

The existence of lines with 100% recurrence, and the grid-like pattern in the colored recurrence plot is evidence of regularities possibly associated with a resilient periodic or quasiperiodic structure that needs to be researched.

In this case, for the theoretical simulation, we find the presence of three 100% recurrence distances corresponding to three cycles, a 21 iterations cycle, a 26 iterations cycle and a 47 iterations cycle. As the total number of iterations is increased and the corresponding 100% recurrence lines are calculated, we find that only these three cycles appear associated with these lines.

The frequencies with which these cycles appear increase linearly with the total number of iterations as can be seen in **Figure 7**, which shows the absolute frequencies of the estimated distances between 100% recurrence lines, estimated for the recurrence matrices with 0.1 radius, with the total number of iterations increasing from 1000 to 10,000 in steps of 1000.

This increase in frequency with the iterations and the multiple distances corresponding to different cycles in the recurrence matrix constitute evidence favorable to the presence of a resilient quasiperiodic structure.

The dominant cycle is 21, followed by the 26-period cycle and, finally, the 47-period cycle. These quasiperiodic signatures are resilient class 2 recurrence signatures that intermix with the class 3 recurrence signatures, in this way, we have evidence of a class 4 recurrence structure, that is, a dynamics that intermixes elements of class 2 and 3.

For the experimental simulation in the quantum device, these cycles also appear, however, other 100% recurrence lines also appear and the cycles are unstable, that is, they last for long sequences of iterations but they are eventually interrupted.

This is linked to two factors: random fluctuations associated with quantum measurement and environmental noise.

In this way, a smaller resilience of these quasiperiodic signatures is expected in the experimental implementation, however, there are a few features that become relevant.

Of the different cycles, there is one that appears often as dominant in the recurrence structure, this is the 47-period cycle, which by contrast is less frequent in the theoretical simulation, this is one of the major differences between the theoretical and the empirical simulation.

Other cycles also appear at some iterations. However, all these cycles are eventually interrupted, and even so, the basic quasiperiodic skeleton resurfaces.

These results are shown in **Figure 8**, which shows, for the empirical simulation in the physical device, the absolute frequencies of the estimated distances between 100% recurrence lines, estimated for the recurrence matrices with 0.1 radius, with the total number of iterations increasing from 1000 to 10,000 in steps of 1000.

Figure 8 contrasts with **Figure 7** on several points, first, besides the 21-, 26- and 47-period cycles there emerges a 68-period cycle, a 115-period cycle and a 162-period cycle, and the 47-period cycle is dominant, instead of the 21 and 26 periods, as stated.

Secondly, these cycles, while present at a certain number of iterations, tend to disappear for a higher number of iterations and then resurface, which means that the 100% recurrence lines, in the physical device, while appearing in iterations with a long number of steps, with the 47-period cycle and the 68-period cycle standing out

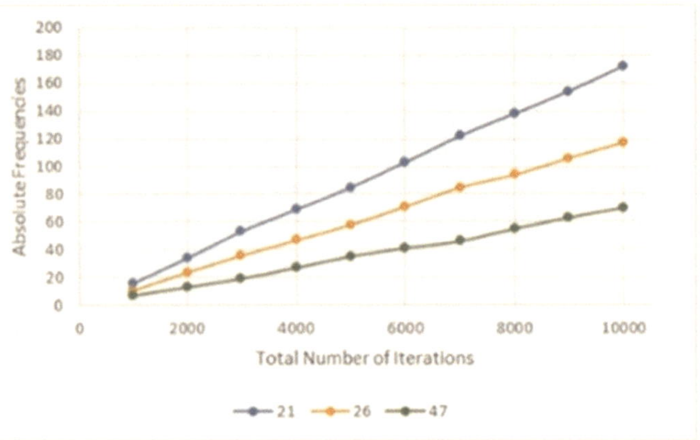

Figure 7.
Statistical distribution of the distances between the 100% recurrence lines for the theoretical simulation with r = 0.550129597 *and the total number of iterations increasing from 1000 to 10,000 in steps of 1000.*

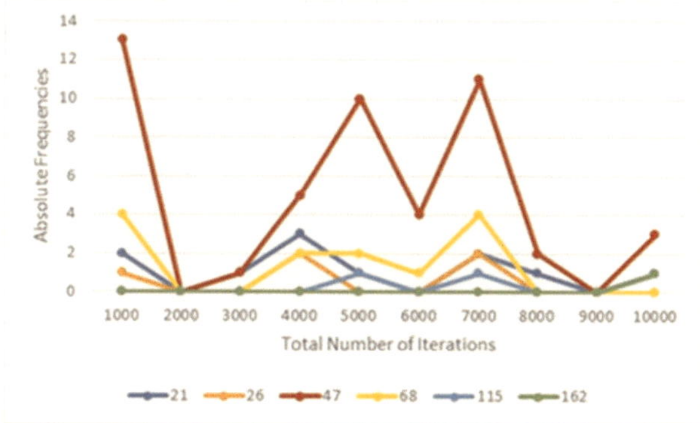

Figure 8.
Statistical distribution of the distances between the 100% recurrence lines for the empirical simulation on "ibm_q_belem" with r = 0.550129597, and the total number of iterations increasing from 1000 to 10,000 in steps of 1000, using 5000 rounds for extracting the quantum averages.

for a 7000 iterations simulation, disappear as the number of iterations is increased which means that the dynamics for the quantum averages is recurrently revisiting the neighborhood of unstable periodic orbits much in the same way as a classical chaotic dynamical system does.

The recurrence structure is driven closer to the topological signatures that also occur in classical chaotic dynamics near the onset of chaos, but already in the chaotic regime.

In the physical device, the dynamics is influenced by the noise and the random fluctuations associated with quantum measurement, leading to fluctuations around the theoretical quantum averages, this factor, and the intrinsically complex mixture of the class 2 and class 3 dynamics, effectively leads to this type of topological signatures that are close to the recurrence signatures of a chaotic dynamics near the onset of chaos.

Now, going beyond the above recurrent network comprised of only two neurons, there is a generalization of this network that intermixes feedforward and recurrent connections.

3. The ring network

The network that we will, now, be addressing is the quantum ring network shown in **Figure 9**. The network is simultaneously a multilayer deep neural network but can also be considered a variant of a quantum cellular automaton.

Defining the binary string s as $s = s_0 s_1 \dots s_{K-1}$, with $s_j \in \{0, 1\}$, for $j = 0, 1, \dots, K-1$, the general state of this network, expanded in the firing pattern basis, is given by the normalized state vector:

$$|\psi> = \sum_s \psi_s |s> \tag{16}$$

With the unitary operators defined as:

$$U_0 = I\left(\otimes_{j=1}^{K-2} I\right) \otimes P_0 + U_r\left(\otimes_{j=1}^{K-2} I\right) \otimes P_1 \tag{17}$$

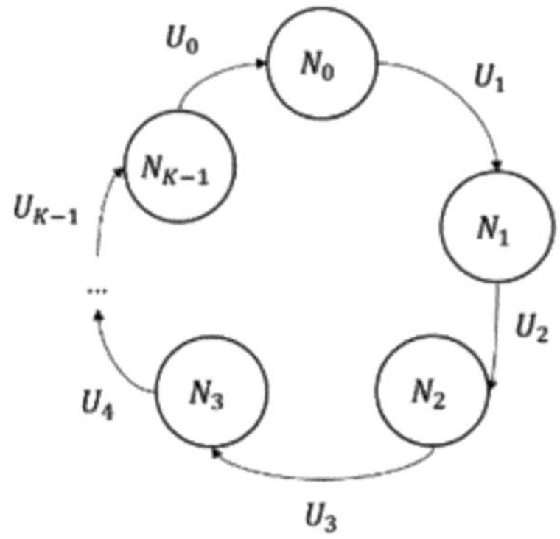

Figure 9.
Quantum ring network.

$$U_l = \left(\otimes_{j=0}^{l-2} I\right) \otimes P_0 \otimes I \left(\otimes_{j=l+1}^{K-1} I\right) + \left(\otimes_{j=0}^{l-2} I\right) \otimes P_1 \otimes U_r \left(\otimes_{j=l+1}^{K-1} I\right), 0 < l < K-1 \tag{18}$$

$$U_{K-1} = \left(\otimes_{j=0}^{K-3} I\right) \otimes P_0 \otimes I + \left(\otimes_{j=0}^{K-3} I\right) \otimes P_1 \otimes U_r \tag{19}$$

The operator U_r is given by Eq. (12) and the unitary map for the network is defined by:

$$F = U_{K-1} \dots U_1 U_0 \tag{20}$$

In the network's simulations, we assume the initial state as:

$$|\psi(0)> = \frac{\sum_s |s>}{\sqrt{2^N}} \tag{21}$$

In this case, instead of the quantum averages for the projectors, we study the local neural firing operators defined as:

$$N_j = \left(\otimes_{j=0}^{j-1} I\right) \otimes |1><1| \left(\otimes_{i=j+1}^{K-1} I\right) \tag{22}$$

Leading to the eigenvalue equation:

$$N_j| \dots s_j \dots > = s_j| \dots s_j \dots > \tag{23}$$

We will be studying, in the simulations, the dynamics of the tuple of quantum averages for these projectors, defined as:

$$N(t) = (\langle\psi(t)|N_0|\psi(t)\rangle, \langle\psi(t)|N_1|\psi(t)\rangle, \dots, \langle\psi(t)|N_{K-1}|\psi(t)\rangle) \tag{24}$$

The above tuple's sequence can be embedded in K-dimensional Euclidean space. In **Figure 10**, we show the dynamics for $r = 0.95$, 5000 iterations and a three-neurons' network, for which we embed the sequences of the quantum operators N_j's averages in three-dimensional Euclidean space, we also show the iterations' sequence.

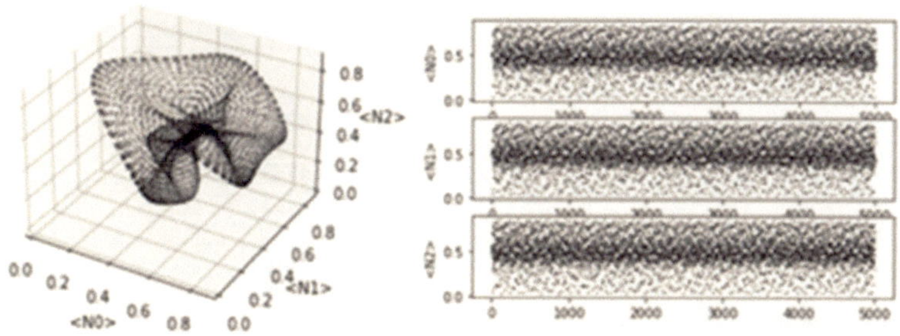

Figure 10.
Embedded dynamics (left) and scatterplot for the local neuron operators' quantum averages (right) for a simulation of the ring network with three neurons, r = 0.95, 5000 *iterations are shown after* 1000 *iterations were dropped for transients.*

Considering the iterations sequence, the dynamics for the local neural firing operators' quantum averages seems to be random-like, however, there is an attractor structure visible in the three-dimensional embedding.

When the number of neurons is increased to four, as shown in **Figure 11**, the dynamics is random and the three-dimensional projection of the four-dimensional attractor is a random scatterplot. In **Figure 12**, we show the colored recurrence plots for each of the two simulations.

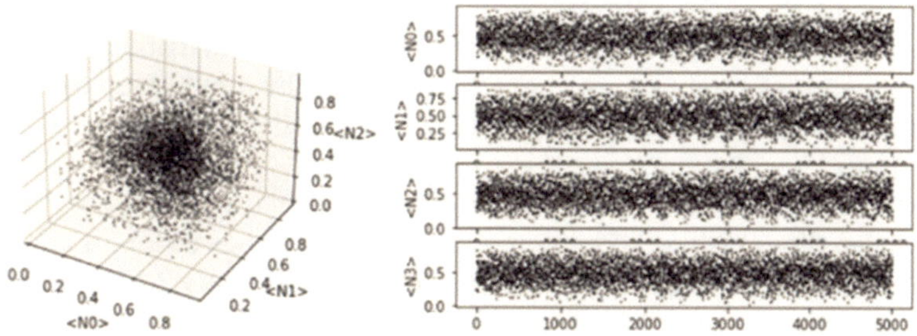

Figure 11.
Three-dimensional projection (left) and scatterplot for the local neuron operators' quantum averages (right) for a simulation of the ring network with four neurons, r = 0.95, 5000 *iterations are shown after* 1000 *iterations were dropped for transients.*

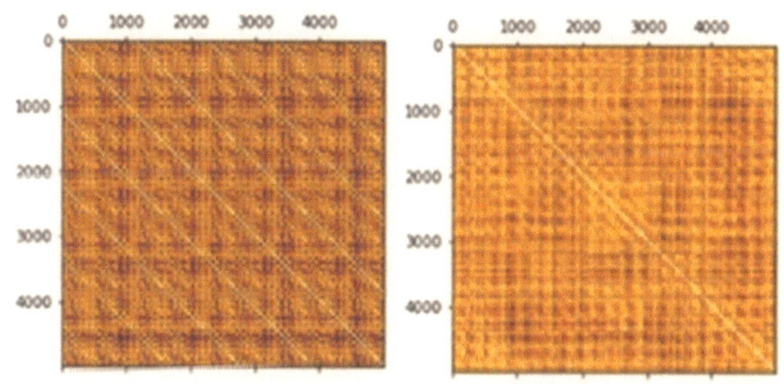

Figure 12.
Colored recurrence plot for the three-neurons' network's simulation (left) and for the four-neurons' network's simulation (right).

Metrics	K = 3	K = 4
Number of lines with 100% recurrence	10	0
Number of lines with recurrence	4769	4747
Number of diagonals	4999	4999
Average recurrence strength	0.0357	0.0056
P[Recurrence]	0.9540	0.9496
P[100%recurrence|recurrence]	0.0021	0

Table 3.
Recurrence metrics for the recurrence matrices for the three- and four-neurons' networks' simulations obtained for a Euclidean radius of 0.1.

Considering the recurrence structure, for the colored recurrence plots and the recurrence metrics for a radius of 0.1 (shown in **Table 3**), we find that the recurrence probability is high for both networks (near 95%), however, the average recurrence strength is low for both networks, which contrasts with the previous results for the two-neurons' network, indicating the emergence of random-like topological signatures that tend to occur in chaotic dynamics.

In this case, the four-neurons' network is close to class 3, since it has high recurrence probability, low average recurrence strength and no line with 100% recurrence, while the three-neurons' network has 10 lines with 100% recurrence, the three-dimensional projection also shows a random-like distribution of points. Since the sequence of quantum averages results from the application of the unitary quantum map, the sequence is actually deterministic, thus, the sequence is actually pseudorandom.

In the case of the three-neurons' network, the dominant part of the recurrence structure is also random-like, indeed, the average recurrence strength is low, which means that, on average, the diagonals with recurrence are only 3.57% occupied with recurrence points and the probability of finding a diagonal with 100% recurrence from a random selection of diagonals with recurrence is just 0.21%, while 95.40% of the diagonals have recurrence points. This means the signatures are also of a predominant class 3 recurrence structure.

Also, for the three-neurons' network, as the number of iterations is increased, unstable cycles emerge, with 100% recurrence lines appearing and then disappearing with the increase in the number of iterations, which are topological features of chaos. This might lead one to consider that the three-neurons' network was a class 3 network with topological markers similar to those of a chaotic dynamics. However, it still has features of class 4 dynamics, with a resilient quasiperiodic skeleton appearing.

Indeed, considering a sequence of simulations from 2000 to 10,000 iterations of this network, in steps of 1000, and calculating the number of cycles, linked to the distances between 100% recurrence lines, for each case, we find that there are unstable cycles appearing as 100% recurrence lines and then disappearing with an increasing number of iterations, this includes a 79-period cycle and a 580-period cycle.

However, there are other cycles that appear as distances between 100% recurrence lines as the number of iterations is increased and that seem to rise with the total number of iterations, these cycles are 299, 388, 499, 617, 728 and 887, **Figure 13** shows these cycles' counts with the iterations increasing from 2000 to 10,000 in steps of 1000. Of notice, all of these resilient cycles tend to increase with the number of

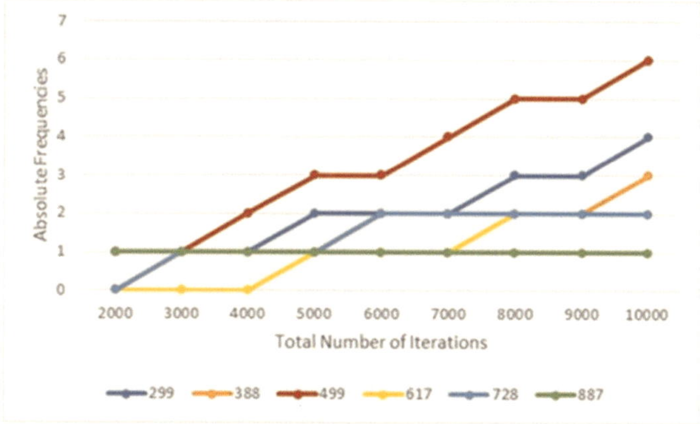

Figure 13.
Persistent cycles associated with distances between 100% recurrence lines for increasing number of iterations for the three-neurons' network, with the number of iterations increasing from 2000 to 10,000 in steps of 1000 and r = 0.95.

iterations in a ladder-like way, except for the 887 cycle which appears only one time and remains as a distance between two diagonals throughout the 10,000 iterations.

Another important point is that the attractor for the three-neurons' network, shown in **Figure 10**, is fractal with a box-counting dimension of 2.1498, using a four-decimal places approximation. **Figure 14** shows the dimension estimation with a fitted line, the resulting R^2 is of 99.7101% and the p-value associated with the slope is 0.0. The box-counting dimension shows that the attractor in **Figure 10** has a dimension that is between 2 and 3 dimensions.

The occurrence of fractal attractors in QNNs' dynamics is not exclusive of the above three-neurons' network, indeed, for the physical simulation of the previously studied two-neurons' network, in the case of 10,000 iterations, the "ibm_q_experience," which, as we saw produced topological markers similar to those of chaos, leads to a form of a sheet that mirrors the triketa shape but with some dispersion, this sheet also has a fractal structure.

In **Figure 15**, we show the resulting attractor along with the estimation of the box-counting dimension, in this case, the estimated dimension is 1.9225, the resulting R^2 is of 99.8504% and the p-value associated with the slope is 0.0.

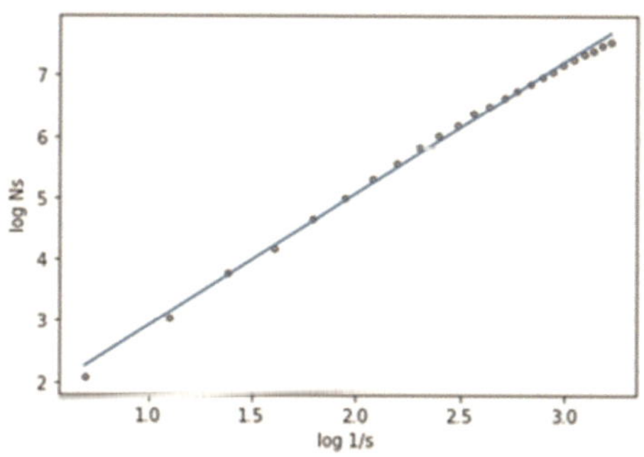

Figure 14.
Estimation of box-counting dimension for the attractor in ***Figure 10.***

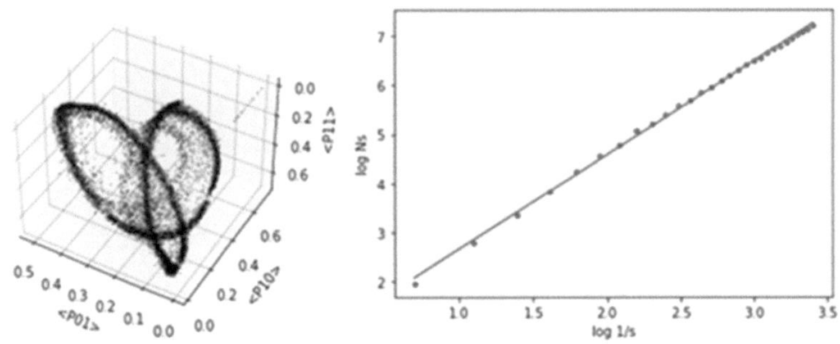

Figure 15.
Simulation of the two-neurons' quantum recurrent neural network on the "ibm_q_belem" device (left), for r = 0.550129597, *with* 5000 *rounds sent for each iteration and* 10,000 *iterations kept after the first* 1000 *iterations were dropped for transients, box-counting dimension estimation (right).*

In this case, unlike the three-neurons' network, which had an estimated fractal dimension between 2 and 3, the estimated fractal dimension for the triketa-like attractor is between 1 and 2, which has the form of a sheet, due to the dispersion induced by noise. In [26] for a quantum stochastic process introduced for this network which included noise in the quantum map also led to a fractal dimension between 1 and 2 for the resulting attractor for the same parameter as the one of **Figure 15**, in that specific case, the estimated dimension was of 1.8927.

As an added note, while most of the points follow the attractor shape, environmental noise fluctuations also led to additional quantum errors that led to the occurrence of a few residual number of points outside the attractor, this can be observed as a straight line that occurred outside the triketa and that is visible in **Figure 15** (left).

These points are not linked to initial transients. They occurred due to noise, corresponding to residual points that are registered in the topological signatures of the quantum averages as random fluctuations that did not lead to a deviation in the scaling of the attractor with the boxes falling close to a straight line in the log-log plot, showing evidence of the fractal scaling.

The fact that the dimension is between 1 and 2 is linked to the noise-like fluctuations which, as can be seen in **Figure 15**, leads actually to a sheet-like structure dispersion of points that follows the triketa, a sheet-like fractal structure as stated also occurred in [26] for the triketa attractor in the case of the stochastic quantum map.

In terms of quantum chaos research, these findings also reinforce other recent work in quantum chaos that also used sequences of averages and embedding methods having identified the emergence of attractors in open quantum systems with the application of topological data analysis methods [29].

The difference, in the case of QNNs, is that one does not start from a classically chaotic system, but rather from a quantum computing network, that is, one is dealing with networked computation.

The emergence of complex attractors in this last case results from the distributed computing dynamics itself. Furthermore, in the case of the unitary map, the dynamics is globally conservative but locally dissipative, that is, each neuron's dynamics is described by a dissipative quantum map.

As studied in detail in [12], for the two-neurons' recurrent network, the entanglement changes with time, leading to local neuron-level von Neumann entropy fluctuations, the study of these fluctuations reveals also complex patterns, that is, the complexity of the network's dynamics extends to the von Neumann entropy dynamics itself. Also, for the global unitary operators, in the case of QNNs, the eigenstates are, in general, entangled states for the networks, which reflects the networked nature of the quantum system in question.

4. Conclusions

We reviewed the application of the concept of a unitary quantum map to a quantum recurrent neural network comprised of two neurons and a QNN with ring topology that includes a feedforward and a recurrent connection, finding, in both cases, complex dynamics with topological signatures typical of chaos and a quasiperiodic resilient skeleton, which characterizes the dynamics as a class 4 recurrence dynamics (edge of chaos).

The quasiperiodic skeleton was also observed in the simulation on IBM's quantum device "ibm_q_experience," but the quasiperiodic signatures are shown to be unstable, and revisited recurrently but occurring as broken long diagonals in the recurrence matrix corresponding to unstable cycles.

Class 3 (random dynamics) and class 2 (regular periodic or quasiperiodic dynamics) were also illustrated. Finally, for both the three-neurons' cycle network and the two-neurons' network's simulation in the quantum device, fractal attractors were shown to be present.

The results reinforce and expand on the previous research into QNNs as complex dynamical systems, also showing how the application of quantum unitary maps and recurrence analysis methods can be effectively employed to study the dynamics of these networks, and how concepts from classical complexity research carry over to quantum complexity research, including the four classes of dynamics that occur for cellular automata, random Boolean networks and coupled map lattices, that can also be identified in the context of recurrence analysis with effectiveness in the characterization of the dynamics of these networks.

In terms of quantum computer science and technologies, these results have implications for the context of quantum distributed computation and the possible dynamical patterns that may occur.

In particular, in the case of class 4 dynamics, the intermixing of chaos-like topological signatures with interrupted diagonals and scattered dots in the recurrence structure, associated with unstable cycles, combined with resilient quasiperiodic signatures means that a form of dynamical memory may be possibly computationally exploited when dealing with quantum computing networks operating as quantum dynamical systems.

On the other hand, the emergence of random-like sequences of quantum averages with the increase in the number of neurons in the ring network demands further study.

From the standpoint of quantum chaos research, these networks may provide for additional models that may lead to complex attractors with fractal signatures and chaos-like recurrence signatures. In this case, the methods of quantum chaos research into unitary and dissipative quantum maps are already shown to be effective when dealing with the dynamics of QNNs.

Author details

Carlos Pedro Gonçalves
Lusophone University, Lisbon, Portugal

*Address all correspondence to: p6186@ulusofona.pt

References

[1] Jensen HJ. Complexity Science: The Study of Emergence. United Kingdom: Cambridge University Press; 2022. 458 p. DOI: 10.1017/9781108873710

[2] Badii R, Politi A. Complexity: Hierarchical Structures and Scaling in Physics. United Kingdom: Cambridge University Press; 1999. 318 p. ISBN: 0 521 66385 7

[3] Nicolis G, Prigogine I. Exploring Complexity: An Introduction. United States: W. H. Freeman; 1989. 313 p. ISBN: 9780716718598

[4] Haken H. Synergetics: Introduction and Advanced Topics. Germany: Springer; 2004. 779 p. ISBN: 978-3540408246

[5] Kaneko K, Tsuda I. Complex Systems: Chaos and beyond. Germany: Springer; 2000. 273 p. DOI: 978-3-642-56861-9

[6] Wolfram S. A New Kind of Science. Canada: Wolfram Media; 2002. 1197 p. ISBN: 978-1579550080

[7] Kauffman SA. The Origins of Order: Self-Organization and Selection in Evolution. United States: Oxford University Press; 1993. 727 p. ISBN: 9780195079517

[8] Arrighi P. An overview of quantum cellular automata. Natural Computing. 2019;**18**:885-899. DOI: 10.1007/s11047-019-09762-6

[9] Franco M, Zapata O, Rosenblueth DA, Gershenson C. Random networks with quantum Boolean functions. Mathematics. 2021; **9**(8):792. DOI: 10.3390/math9080792

[10] Wójcik DK. Quantum maps with space extent: A paradigm for lattice quantum walks. International Journal of Modern Physics B. 2006; **11**:34. DOI: 10.1142/S02179792 06034509

[11] Grosvenor KT, Jefferson R. The Edge of Chaos: Quantum Field Theory and Deep Neural Networks [Internet]. 2022. Available from: https://arxiv.org/pdf/2109.13247.pdf [Accessed: October 17, 2023]

[12] Gonçalves CP. Quantum neural networks, computational field theory and dynamics. International Journal of Swarm Intelligence and Evolutionary Computation. 2022;**11**:245. DOI: 10.35248/2090-4908.22.11.246

[13] Prigogine I. From classical chaos to quantum chaos. Vistas in Astronomy. 1993;**37**:7-25. DOI: 10.1016/0083-6656(93)90005-5

[14] Reichl L. The Transition to Chaos: Conservative Classical and Quantum Systems. Switzerland: Springer; 2021. 555 p. DOI: 10.1007/978-3-030-63534-3

[15] Hummel Q, Richter K, Schlagheck P. Genuine many-body quantum scars along unstable modes in Bose-Hubbard Systems. Physical Review Letters. 2023;**130**:250402. DOI: 10.1103/PhysRevLett.130.250402

[16] Tomsovic S, Urbina JD, Richter K. Controlling quantum chaos: Optimal coherent targeting. Physical Review Letters. 2023;**130**:020201. DOI: 10.1103/PhysRevLett.130.020201

[17] Tomsovic S, Urbina JD, Richter K. Controlling quantum chaos: Time-dependent kicked rotor. Physical Review E. 2023;**108**:044202. DOI: 10.1103/PhysRevE.108.044202

[18] Menneer T. Quantum Artificial Neural Networks [thesis]. Exeter: The University of Exeter; 1998

[19] Narayanan A, Meneer T. Quantum artificial neural network architectures and components. Information Sciences. 2000;**128**:231-255. DOI: 10.1016/S0020-0255(00)00055-4

[20] Schuld M, Sinayskiy I, Petruccione F. The quest for a quantum neural network. Quantum Information Processing. 2014;**13**(11):2567-2586. DOI: 10.1007/s11128-014-0809-8

[21] Gonçalves CP. Quantum cybernetics and complex quantum systems science: A quantum connectionist exploration. NeuroQuantology. 2015;**13**(1):35-48. DOI: 10.14704/nq.2015.13.1.804

[22] Gonçalves CP. Quantum neural machine learning: Backpropagation and dynamics. NeuroQuantology. 2017; **15**(1):22-41. DOI: 10.14704/nq.2017.15.1.1008

[23] Braun D. Dissipative Quantum Chaos and Decoherence. Germany: Springer; 2001. 132 p. DOI: 10.1007/3-540-40916-5

[24] Packard NH. Adaptation toward the edge of chaos. Dynamic Patterns in Complex Systems. 1988;**212**:293-301. DOI: 10.1142/9789814542043

[25] Langton CG. Computation at the edge of chaos: Phase transitions and emergent computation. Physica D: Nonlinear Phenomena. 1990;**42**(1-3): 12-37. DOI: 10.1016/0167-2789(90)90064-V

[26] Gonçalves CP. Quantum stochastic neural maps and quantum neural networks. SSRN Research Paper. 2019: 3502121. DOI: 10.2139/ssrn.3502121

[27] Eckmann JP, Kamphorst SO, Ruelle D. Recurrence plots of dynamical systems. Europhysics Letters. 1987;**4**(9): 973-977. DOI: 10.1209/0295-5075/4/9/004

[28] Gao J, Cai H. On the structures and quantification of recurrence plots. Physics Letters. 2000;**270**:75-87. DOI: 10.1016/S0375-9601(00)00304-2

[29] Cao H, Leykam D, Angelakis DG. Unravelling quantum chaos using persistent homology. Physical Review E. 2023;**107**:044204. DOI: PhysRevE.107.044204

[30] Gardiner CW, Zoller P. Quantum Noise – A Handbook of Markovian and Non-Markovian Quantum Stochastic Methods with Applications to Quantum Optics. Germany: Springer; 2004. 449 p. ISBN: 3-540-22301-0

[31] Gonçalves CP. NSimul.ipynb [Internet]. 2023. Available from: https://github.com/cpgoncalves/qneural/blob/master/NSimul.ipynb

[32] Gonçalves CP. QNeural [Internet]. 2019. Available from: https://github.com/cpgoncalves/qneural

Chapter 7

An Alternative Approach to Probability in Quantum Information Science

Christian Jansson

Abstract

This chapter provides a probabilistic framework for formulating classical probability theory, quantum probability, thermodynamics, diffusion, and the Wiener integral using a set of four axioms or principles. It explains everything that conventional quantum information theory and classical probability theory achieve. We want to emphasize that this framework is not an interpretation of quantum mechanics such as "Many-Worlds," "Bohm's Theory," or the "Copenhagen interpretation." It is much more general and can be viewed as a probability algorithm that calculates probabilities of future events. As a result, previously perplexing paradoxes find resolution. In particular, the superposition principle takes on a new meaning. Our probabilistic framework stands apart from the Hilbert space formalism. It relies solely on elementary set theory, classical logic, and complex numbers. Consequently, this theory is accessible for instruction in educational settings. This framework can be regarded as an axiomatic approach to probability in the sense of Hilbert. In his sixth of the twenty-three open problems presented at the International Congress of Mathematicians in Paris in 1900, Hilbert called for an axiomatic probability theory.

Keywords: quantum information theory, quantum physics, probability axioms, Feynman's formulation, thermodynamics, diffusion, Brownian motion

1. Introduction

The true logic of the world is in the calculus of probabilities. James Clerk Maxwell

According to the Cambridge dictionary, probability is a nonnegative number representing how likely a particular outcome in a random experiment will happen. David Hilbert introduced his renowned set of fundamental problems, including the sixth problem, which aimed to establish an axiomatic foundation for probability theory, much like geometry. Several notable responses to Hilbert's challenge have resurfaced since then, see [1]. However, it is worth noting that more than a century later, von Weizsäcker ([2], p. 59), further delved into this topic:

Probability is one of the outstanding examples of the epistemological paradox that we can successfully use our basic concepts without actually understanding them. von Weizsäcker [2]

Even today, classical probability, its axioms, and how to assign probabilities to elementary events is a philosophical dispute discussion. There are various interpretations of probability, including one of the oldest, the frequency interpretation. See [1, 3], and the literature therein for different interpretations and axioms.

The concepts of information and probability are closely linked. For instance, the Shannon concept of information relies on probability, such as the thermodynamic concept of entropy. In the same way, quantum states are probabilistic states of information.

About the probability in quantum information science, Weinberg [4] writes:

Even so, I'm not as sure as I once was about the future of quantum mechanics. It is a bad sign that those physicists today who are most comfortable with quantum mechanics do not agree with one another about what it all means. The dispute arises chiefly regarding the nature of measurement in quantum mechanics. Weinberg [4]

Regarding quantum probability problems, discussions often take on a strange and peculiar attitude, as Fuchs [5] noted in his reflections on the annual conferences.

What is the cause of this year-after-year sacrifice to the "great mystery?" Whatever it is, it cannot be for want of a self-ordained solution: Go to any meeting, and it is like being in a holy city in great tumult. You will find all the religions with all their priests pitted in holy war — the Bohmians [3], *the Consistent Historians* [4], *the Transactionalists* [5], *the Spontaneous Collapseans* [6], *the Einselectionists* [7], *the Contextual Objectivists* [8], *the outright Everettics* [9, 10], *and many more beyond that. They all declare to see the light, the ultimate light. Each tells us that if we accept their solution as our savior, then we too will see the light.* Fuchs [5]

A detailed description of interpretations of quantum mechanics, including many references, can be found in [6]. We also mention the easily readable WIKIPEDIA article "interpretations of quantum mechanics."

In this chapter, we introduce a predictive algorithm designed for calculating probabilities concerning future macroscopic events or detector clicks. Our principles maintain a strict separation of internal possibilities and outcomes, leading to a broader scope than the axioms of quantum mechanics. This chapter summarizes some essential parts of three lecture notes [7–10], including some corrections.

The chapter is organized as follows. The basic axioms of classical probability and the fundamental add and multiply rule, meaning that "probabilities for disjoint events are added, and probabilities for independent events are multiplied," are introduced in Section 2. Moreover, it is emphasized that classical probability and quantum probability are not compatible. The main topic of this chapter is a probabilistic framework, consisting of four general principles, which, in particular, allow the reconstruction of classical probability *and* quantum probability. These principles form the content of Section 3. Several examples, including the double-slit experiment, are presented in Section 4. Some more details about Hilbert's sixth problem are given in Section 5. In Section 6, we show that our probabilistic framework is consistent and contains a $U(1)$ symmetry as in quantum electrodynamics. In Section 7, we present the reconstruction of Feynman's formulation of quantum mechanics, which is mathematically equivalent to Schrödinger's theory and Heisenberg's matrix mechanics. Its close relationships to Brownian motion, Wiener integrals, and diffusion are described in Section 8. The reconstruction of statistical thermodynamics, described in Section 9, is a vital touchstone when applying a probabilistic theory. Finally, some conclusions are given.

2. The Kolmogorov axioms

In 1933, Kolmogorov presented a mathematical theory of classical probability in terms of some axioms that have since become standard. He uses elementary set theory. Given two sets A and B, the *union* $A \bigcup B$ denotes the set whose elements appear either in A or in B, or in both. The *intersection* $A \cap B$ denotes the set whose elements appear in both sets A and B. If A is a subset of B, then the *complement* A^c is the subset of B whose elements are not in A.

Kolmogorov's axioms are based on two fundamental sets:

i. The *sample space* $\mathbf{O}$ of outcomes. The outcomes are also called *elementary events*.

ii. The *probability algebra* $\mathbf{A}$ of subsets of $\mathbf{O}$ that contains $\mathbf{O}$ itself and is closed under complement and countable unions. Sometimes, this algebra is called *field*. The subsets $A \in \mathbf{A}$ are called *events*.

Moreover, there exists a mapping Pr, called *probability* from the field of events $A \in \mathbf{A}$ into the set of nonnegative numbers:

iii.

$$A \to \Pr(A), \quad 0 \leq \Pr(A) \leq 1, \quad \Pr(\mathbf{O}) = 1, \tag{1}$$

and for any countable set of disjoint events A_m, the equation

$$\Pr\left(\bigcup_{m=1}^{\infty} A_m\right) = \sum_{m=1}^{\infty} \Pr(A_m). \tag{2}$$

must be fulfilled.

The condition that probabilities are numbers between zero and one is essential since otherwise, we cannot hope that the relative frequencies of an event A approach $\mathrm{PR}(A)$. The *relative frequency* is the number of times the event A occurred in a series of executions of an experiment divided by the number of executions, thus being bounded between zero and one.

Two events A and B are called *independent* if both do not influence each other. For instance, if we toss a coin twice and know the outcome A of the first toss, then this has no influence on the result B of the second toss. The probabilities for independent events are multiplied, that is,

$$\Pr(A \cap B) = \Pr(A)\Pr(B) \tag{3}$$

In summary, the probabilities of disjoint events are added, and the probabilities of independent events are multiplied. This is the well-known *multiply and add rule*, which holds valid already for Laplace experiments.

The mathematics of Kolmogov's probability theory is well understood, but its interpretation is controversial, see also [3].

It is well-known that classical probability and quantum probability are incompatible and contradictory. For example, Anthony Zee [11] writes on page 141 under the title "Dice Unlike Any Dice:"

> *Welcome to the strange world of the quantum, where one cannot determine how a particle gets from here to here. [...] When a die is thrown, the probability of getting a 1 is 1/6. The probability of getting a 2 is, of course, also 1/6. Now, consider the following question: What is the probability of getting a 1 or a 2 in one throw? The answer is evident to gamblers and non-gamblers alike: The probability is 1/6 + 1/6 =1/3. In everyday life, to obtain the probability of either A or B occurring, we simply add the probability of A occurring and the probability of B occurring.*
>
> *The quantum die is astonishingly different. Suppose we are told that for the quantum die the probability of throwing a 1 is 1/6, and the probability of throwing a 2 is also 1/6. In contrast to what our experience with ordinary dice might suggest, we cannot conclude that the probability of getting either a 1 or a 2 in one throw is 1/3! It turns out that the probability of throwing a 1 or a 2 can range between 1/3 and 0!*

Apparently, quantum theory yields results other than classical probability theory, and the question arises about a more fundamental theory below both theories.

The main aim of this publication is to present a general probability theory that simultaneously allows the treatment of classical stochastic and quantum mechanical experiments.

3. A unified probabilistic framework

An opinion on quantum mechanics, held by numerous physicists, is eloquently articulated in the book authored by Susskind and Friedman ([12], p. 24):

> *For a classical system, the space of states is a set (the set of possible states), and the logic of classical physics is Boolean. That seems obvious, and it is not easy to imagine any other possibility. Nevertheless, the real world operates along different lines, at least whenever quantum mechanics is important. The space of states of a quantum system is not a mathematical set [6]; it is a vector space. Relations between the elements of a vector space are different from those between the elements of a set, and the logic of propositions is different as well.*

Is a Hilbert space formalism and a modified logic indispensable in quantum physics? We describe a probabilistic framework that unifies classical mechanics, statistical thermodynamics, and quantum mechanics, not based on Hilbert spaces but on classical logic. It is based on decidable alternatives, which we call *outcomes*. The outcomes are described by sets consisting of *elementary possibilities*. Our approach partially supports the opinion of Fuchs and Peres [13]:

> *The thread common to all the nonstandard "interpretations" is the desire to create a new theory with features corresponding to some reality independent of our potential experiments. But, trying to fulfill a classical worldview by encumbering quantum*

> *mechanics with hidden variables, multiple worlds, consistency rules, or spontaneous collapse without any improvement in its predictive power only gives the illusion of a better understanding. Contrary to those desires, quantum theory does not describe physical reality. What it does is provide an algorithm for computing probabilities for the macroscopic events ("detector clicks") that are the consequences of our experimental interventions. This strict definition of the scope of quantum theory is the only interpretation ever needed, whether by experimenters or theorists.* Fuchs and Peres [13]

In the following, we consider random experiments in the broadest sense. They are described by three sets:

i. The *possibility space* $\mathbf{P}$ is a set with elements $p \in \mathbf{P}$. We call its elements *elementary possibilities*.

ii. The *possibility algebra* $\mathbf{F}$ is defined as the collection of subsets of $\mathbf{P}$ that contains $\mathbf{P}$ itself and is closed under complement and countable unions, that is, $\mathbf{F}$ is a *field*. The subsets $F \in \mathbf{F}$ are called *possibilities*. If F does not correspond to an elementary possibilities $\{p\}$, then F is called *nonelementary*.

iii. The *sample space* $\mathbf{O}$ consists of pairwise disjoint sets $F \in \mathbf{F}$ called *outcomes*. The outcomes form a partition of the possibility space, that is, each elementary possibility $p \in \mathbf{P}$ is contained in exactly one outcome F. If an outcome F consists of more than one element, we call its elements *internal elementary possibilities*, which are *accessible* from F.

   Additionally, we demand the existence of a function that evaluates possibilities:

iv. A *probability amplitude* is defined as a mapping φ from the possibility algebra $\mathbf{F}$ into the set of complex numbers:

$$F \to \varphi_F = \varphi(F) \in \mathbb{C},\ F \in \mathbf{F}. \tag{4}$$

We claim that these quantities satisfy two general principles or axioms.

First principle: Given a countable set of pairwise disjoint possibilities $F_m \in \mathbf{F}$, in order that $F = \bigcup_m F_m$, it is

$$\varphi_F = \varphi\left(\bigcup_m F_m\right) = \sum_m \varphi_{F_m}. \tag{5}$$

This principle is the *superposition of probability amplitudes*. It is very general compared to Feynman's first principle: "When an event can occur in several alternative ways, the probability amplitude for the event is the sum of the probability amplitudes for each way considered separately" ([14], pp. 1–16). Feynman's quantum mechanics does not distinguish between outcomes, possibilities, and internal possibilities. Hence, it differs from our framework. Moreover, Feynman uses the four-dimensional space-time, and we require only the partitioning future, present, and past.

Second principle: This is *Born's rule*. It transforms probability amplitudes of outcomes F to probabilities $Pr(F)$:

$$\Pr(F) = |\varphi_F|^2 \text{ for all } F \in \mathbf{O}, \text{ and } \sum_{F \in \mathbf{O}} |\varphi_F|^2 = 1. \tag{6}$$

Born's rule states that the probability of measuring a specific outcome F is proportional to the square of the absolute value of the probability amplitude associated with that outcome. Summing up the probabilities of all outcomes, we get one, such that Born's rule implies a real probability measure on the sample space $\mathbf{O}$. In particular, classical probability is incorporated.

We call the quadruplet $(\mathbf{P}, \mathbf{F}, \mathbf{O}, \varphi)$, together with these two principles, a *possibility measure space*.

It is worth noting that, in the literature, a measure is often defined as a nonnegative function, in contrast to the use of complex amplitudes here. Nevertheless, it is crucial to emphasize that complex numbers are both indispensable and fundamental for accurately describing quantum physical reality, as supported by several references: [12], page 44, [7], Section 2.2, [15].

The probabilities for the outcomes belong to the prognostic category future. In the category present, the experiment is performed. For example, a particle runs through the experimental setup. This particle has no idea of the experimental setup and the placed detectors. The only thing it does is to act according to the probabilities: There is no rest, and the particle tends to move toward states of larger probability. Notice that this point of view is fundamentally different compared to the widely celebrated wave-particle duality.

These two principles are mathematical conditions that must be satisfied for probability amplitudes. The first principle implies that it is sufficient to compute the amplitudes for the elementary possibilities only. The second one, Born's rule, says we must calculate only the amplitudes for the outcomes. Now, we introduce two further principles that help to compute concrete probability amplitudes.

Third principle: The amplitudes φ_F contribute equally in magnitude for all accessible elementary possibilities. They are proportional to some constant times a complex number of magnitude one, namely

$$e^{\frac{i}{\hbar}S(F)}. \tag{7}$$

The function $S(F)$ is called the *action* of the elementary possibility F.

Feynman's formulation of quantum theory is as follows:

> *The total amplitude can be written as the sum of amplitudes of each path - for each way of arrival. For every $x(t)$ that we could have - for every possible imaginary trajectory - we have to calculate an amplitude. Then, we add them all together. What do we take for the amplitude of each path? Our action integral tells us what the amplitude for a single path ought to be. The amplitude is proportional to some constant times $\exp(iS/\hbar)$, where S is the action for the path. If we represent the phase of the amplitude by a complex number, Planck's constant $\hbar$ has the same dimensions.*
> Feynman and Hibbs [16], page 19.

In our third principle, no further requirements are made about the action, as is necessary in the case of space-time paths. Consequently, this principle is very flexible in describing physical problems outside space-time.

In classical probability theory, the Laplace *principle of indifference* says that all outcomes are equally likely assigned with unit one. Hence, the difference is that we merely replace unit one with complex numbers of magnitude one. It follows that we get back the theory of Laplace if we set the phase equal to zero.

Fourth principle: We call two possibilities F and G *independent* provided their intersection is non-empty, and the occurrence of one possibility does not affect the other one. Mathematically, independence is defined by the equation:

$$\varphi_{F\cap G} = \varphi_F \ \varphi_G. \tag{8}$$

In other words, the joint amplitude is equal to the product of their amplitudes.

Feynman ([14], pp. 3–4), describes independence as follows: "When a particle goes by some particular route, the amplitude for that route can be written as the product of the amplitude to go partway with the amplitude to go the rest of the way." Independence is a fundamental concept in probability theory and statistics because it simplifies calculations and allows for modeling complicated random experiments. Laplace already introduced it in the late eighteenth and early nineteenth centuries. It says that an experiment, which breaks down into a series of possibilities happening independently, the probability of the occurrence of all possibilities is the product of the probability of each. Our first and fourth principle shows that the well-known *multiply and add rule* in probability theory carries over to complex probability amplitudes for possibilities.

The physical content of this theory lies in the third principle *via* the classical action $S(F)$. In contrast, the other principles are purely mathematical and physically empty. Notice that the stationary-action principle is a variational principle, yielding the equations of motion in Newtonian mechanics, general relativity, and classical electrodynamics when applied to the corresponding action. For example, in general relativity, it is the Einstein-Hilbert action. In quantum field theory, the action is incorporated into the path integral.

4. Examples

4.1 Tossing a die

A simple example is tossing a fair die. There are six elementary possibilities $1, 2, 3, 4, 5, 6$ yielding the possibility space $\mathbf{P} = \{1, 2, 3, 4, 5, 6\}$. The corresponding possibility algebra $\mathbf{F}$ is the power set of $\mathbf{P}$. The six outcomes correspond to the six elementary possibilities. They form a partition of the possibility space. We define the action as equal to zero. Then, the exponential interference term in Eq. (7) is equal to one. We set

$$\varphi_\varnothing = 0, \ \ \varphi_{\{k\}} = \frac{1}{\sqrt{6}}, \ \ \text{for all} \ \ k \in \mathbf{P}. \tag{9}$$

Hence, the probabilities for all outcomes are $1/6$, according to the second principle. The first principle yields

$$\varphi_{\{\mathbf{P}\}} = 6 \times \frac{1}{\sqrt{6}} = \sqrt{6}. \tag{10}$$

The value $\left|\varphi_{\{\mathbf{P}\}}\right|^2 > 1$ is no contradiction because Born's rule is only applied to the outcomes, not to arbitrary possibilities. Notice that the probabilities of the outcomes satisfy Kolmogorov's axioms.

The probability amplitudes and the related probabilities are prognostic numbers for future events. If we execute an experiment in the present, the results agree with these probabilities.

4.2 Atom in two states

In his book, Smolin [17], Chapter 4, presents a simple quantum experiment of an atom that can exist in two states: an excited state denoted as E and a ground state with the lowest energy, denoted as G. The atom, while in the unstable excited state E, has the capability to transit to the ground state G by emitting a photon. To explore this, we place an excited atom inside a sealed box alongside a Geiger counter. Much like the atom, the Geiger counter has two possible states: the yes-state Y, meaning that it has detected a photon, and the no-state N indicating that no photon has been detected.

Initially, the system is in the state (E, N) with the atom in the excited state and the Geiger counter in the no state. After a certain duration, when we open the box, we find that the system is in one of two possible states: Either it remains in the initial state (E, N) or it has transitioned to the state (G, Y) with the atom in the ground state and the Geiger counter in the yes state.

The postulates of quantum mechanics tell us that, before opening the box, the system is in a *superposition* of both states

$$(E, N) \ \textbf{and} \ (N, Y). \tag{11}$$

Smolin calls this the "in-between" state. However, we have never observed a superposition after opening the box. This is a seemingly weird situation, as Smolin writes. Then, he raises some questions. Why has quantum mechanics two dynamical rules, the *unitary evolution* before opening the box, and the *collapse* into one of the states (E, N) *or* (G, Y) when opening the box? This is in contrast to other theories that only have one dynamic. Why does the process of measurement differ from other processes? When does the collapse occur? Does it happen when the particle interacts with the counter or when the box is opened, and an observer becomes conscious of the outcome? These and other questions are typical in quantum mechanics.

Our framework is purely probabilistic, calculating numbers of future events. In the present, when performing an experiment, the detector clicks agree with the calculated probabilities in the sense of relative frequencies. Questions as above do not occur. For the atom with two states and the Geiger counter, the possibility space has the form

$$\mathbf{P} = \{(E, N), (E, Y), (G, N), (G, Y)\}. \tag{12}$$

The related possibility algebra is the power set of $\mathbf{P}$. The outcomes coincide with the elementary possibilities. In other words, the sample space of outcomes and the possibility space are identical if we identify $p \in \mathbf{P}$ with $\{p\} \in \mathbf{F}$. There exist no internal possibilities. We have a simple classical statistical situation.

We set

$$\varphi(E, Y) = \varphi(G, N) = 0, \quad \varphi(E, N) \neq 0, \quad \varphi(G, Y) \neq 0. \tag{13}$$

All quantities belong to the prognostic future, that is, they describe what might happen. The nonzero amplitudes depend on the experimental setup, the type of atoms, and how long the particle is in the closed box. Born's rule tells us that

$$|\varphi(E, N)|^2 + |\varphi(G, Y)|^2 = 1. \tag{14}$$

In the present, the experimental results show that the atom tends to move to higher probability outcomes. Nothing is strange; we require no "in-between" superpositions.

Smolin's example is directly related to the well-known Schrödinger's cat thought experiment, a famous quantum mechanical illustration devised by Schrödinger in 1935. It is designed to highlight the concept of superposition or "in-between" states and the peculiar nature of quantum mechanics. In the experiment, a hypothetical cat is placed in a sealed box with a radioactive atom, a Geiger counter, a vial of poison, and a mechanism that will release the poison if the Geiger counter detects radiation. According to the principles of quantum mechanics, before the box is opened and the cat is observed, the cat's state is in between alive and dead. In other words, until the box is opened, the cat is considered alive and dead simultaneously. In our framework, the cat is either dead or alive. Nothing strange happens.

4.3 The double-slit experiment

The double-slit experiment has been called "the most beautiful experiment in physics" [18]. Frequently, its interpretation is that particles of matter behave like a wave and that the act of observing a particle has a dramatic effect on the experimental results. It can be performed with photons or electrons. Actually, experiments with large molecules composed of more than 800 atoms indeed show interference. In 2012, physicists from Vienna used large molecules called phthalocyanine, which can be seen with a video camera exhibiting their macroscopic nature. The experiment is executed such that only one molecule interacts with the setup. They arrive localized at small places at the final wall of detectors, a behavior typical for macroscopic objects, not for classical waves.

Imagine a source producing particles. Behind is a wall with two slits in it and, after that, a screen of detectors. If we execute the experiment, some particles will bounce off the wall, but some will travel through the slits and will arrive at the screen, see **Figure 1**. We consider only the particles arriving at the screen. We define the elementary possibilities as follows: They are piecewise straight paths sad_m, sbd_m from the source s, *via* the wall W with two slits a and b, to the detectors $d_m, m = -l, \ldots, l$ positioned on the screen D.

We consider three experimental setups. Firstly, only one slit is open. Secondly, both slits are open, and thirdly detectors measure through which slit the particle goes. We shall see how these changes in the experimental setup change the statistics significantly.

Firstly, let slit b be closed, that is, only paths through slit a are relevant. Hence, the possibility space is

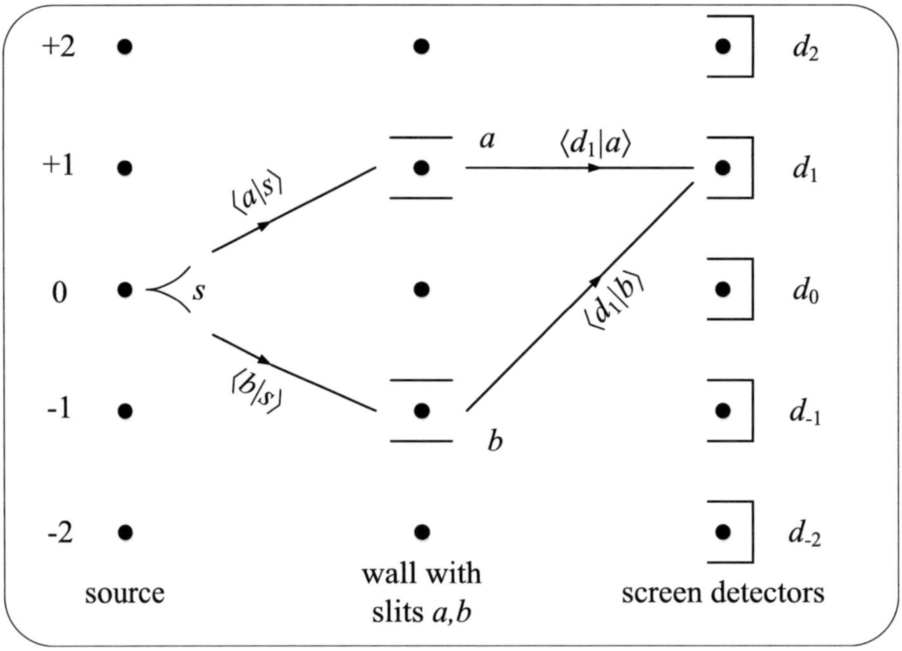

Figure 1.
The double-slit experiment: a particle leaves source s, passes one of the two slits a or b, and is detected in d_1, finally.

$$\mathbf{P} = \{sad_m : \ d_m \in D\}. \tag{15}$$

We have no internal possibilities such that the sample space of outcomes

$$\mathbf{O} = \{O_{d_m} : \ d_m \in D\}, \ O_{d_m} = \{sad_m\} \in \mathbf{F} \tag{16}$$

corresponds uniquely to **P**. This is a classical experiment. Our third principle yields the amplitude

$$\varphi(O_{d_m}) = \varphi_{sad_m} \tag{17}$$

via the action of the path sad_m. The squared magnitudes of the amplitudes are the probabilities:

$$\Pr(O_{d_m}) = \ \left|\varphi_{sad_m}\right|^2 \tag{18}$$

In the same way, we obtain the probability

$$\Pr(O_{d_m}) = \ \left|\varphi_{sbd_m}\right|^2, \tag{19}$$

when slit a is closed.

Now secondly, we suppose that both slits are open. Then the possibility space consists of all paths from the source to the detectors

$$\mathbf{P} = \{sad_m, sbd_m : \ a, b \in W, d_m \in D\}. \tag{20}$$

We have internal possibilities since it cannot be observed through which slit the particle goes in the present. The sample space of outcomes is defined as

$$\mathbf{O} = \{O_{d_m} : \ d_m \in D\}, \quad \text{where} \ \ O_{d_m} = \{sad_m, sbd_m\}. \tag{21}$$

Using the third principle, we set

$$\varphi_{sad_m} = \frac{1}{\sqrt{2}} e^{\frac{i}{\hbar}S(sadm)}, \ \ \varphi_{sbd_m} = \frac{1}{\sqrt{2}} e^{\frac{i}{\hbar}S(sbdm)}, \tag{22}$$

and the first principle yields the amplitudes of the outcomes

$$\varphi(\{O_{d_m}\}) = \varphi_{sad_m} + \varphi_{sbd_m} \ \ \text{for all} \ \ d_m \in D. \tag{23}$$

Born's rule provides the probabilities of the outcomes:

$$\begin{aligned}
\Pr(O_{d_m}) &= \left| \frac{1}{\sqrt{2}} e^{\frac{i}{\hbar}S(sadm)} + \frac{1}{\sqrt{2}} e^{\frac{i}{\hbar}S(sbdm)} \right|^2 \\
&= \frac{1}{2}\left(\left| e^{\frac{i}{\hbar}S(sadm)} \right|^2 + \left| e^{\frac{i}{\hbar}S(sbdm)} \right|^2 \right) \\
&+ \frac{1}{2}\left(\left(e^{\frac{i}{\hbar}S(sadm)} \right)^* e^{\frac{i}{\hbar}S(sbdm)} + \left(e^{\frac{i}{\hbar}S(sbdm)} \right)^* e^{\frac{i}{\hbar}S(sadm)} \right).
\end{aligned} \tag{24}$$

Compared with the case where one slit is closed, the first term in this sum corresponds to the classical probability. The second term is responsible for interference.

In the case where $e^{\frac{i}{\hbar}S(sadm)} = e^{\frac{i}{\hbar}S(sbdm)}$, Eq. (24) yields

$$\Pr(O_{d_m}) = 2\left| e^{\frac{i}{\hbar}S(sadm)} \right|^2. \tag{25}$$

Thus, the probability when only one slit is open is doubled, and we get *constructive interference*. For the other extreme case where $e^{\frac{i}{\hbar}S(sadm)} = -e^{\frac{i}{\hbar}S(sbdm)}$, we get the probability

$$\Pr(O_{d_m}) = 0, \tag{26}$$

yielding *destructive interference*.

We have computed only probabilities of future events, yielding a pattern of constructive and destructive interference. In the present, a particle chooses a path. Preferably, those with high probability.

Finally, suppose we have information about the slit where a particle passes. This information comes about by two additional detectors d_a and d_b at the slits. Assume that the detectors work correctly such that it cannot happen that a particle arrives at detector d_m *via* slit b and detector d_a clicks, or both detectors d_a and d_b do not click.

Then the possibility space is defined as

$$\mathbf{P} = \{sad_a d_m, sbd_b d_m : \ \ a, b \in W, d_m \in D\}. \tag{27}$$

The outcomes are defined via the detector clicks at the screen and the clicks of two additional detectors d_a and d_b. Hence, we obtain the sample space of outcomes:

$$\mathbf{O} = \{O_{d_a d_m}, O_{d_b d_m} : d_m \in D\}, \tag{28}$$

where

$$O_{d_a d_m} = \{sad_a d_m\}, O_{d_b d_m} = \{sbd_b d_m\}. \tag{29}$$

Using Born's rule, we get the classical probabilities:

$$\Pr(O_{d_a d_m}) = |\varphi_{sad_m}|^2, \ \Pr(O_{d_b d_m}) = |\varphi_{sbd_m}|^2. \tag{30}$$

In summary, the numbers computed for the three different experimental setups are probabilities that describe how likely in the *future* a particle would meet one of the detectors. In the *present*, the particle does not know anything about the experimental setup. It passes the experiment with the tendency to move on exactly one path of higher probability. Of course, in rare cases, the particle will also choose paths with low probability. This natural explanation is all what we need to know. We see that it is essential to distinguish clearly between *elementary possibilities* and *outcomes*. Then interpretations, such as "wave-particle dualism," "many-worlds," "non-locality," and others are unnecessary. In particular, a material object does not occupy several locations at the same time as Penrose writes in his excellently written book [19] on page 216:

> *As we have seen, particularly in the previous chapter, the world actually does conspire to behave in a most fantastical way when examined at a tiny level at which quantum phenomena hold sway. A single material object can occupy several locations at the same time and like some vampire of fiction (able, at will, to transform between a bat and a man) can behave as a wave or as a particle seemingly as it chooses, its behavior is governed by mysterious numbers involving the "imaginary" square root of -1.* Penrose [19]

Penrose gave, not unfounded, his book the title FASHION, FAITH, and FANTASY. Our aim is, however, to show that the world is stochastic, at least their physical descriptions, but in no way fantastical and mysterious.

Large macroscopic molecules or other objects can be described as a cloud of elementary particles the constituents. Suppose the binding force between these constituents is very weak. Then the constituents in this cloud can independently of one another move through both slits yielding interference. But when the binding force between the constituents is large, then all move through the same slit. Then we get a stochastic pattern as in the case where only one slit is open.

It is easy to generalize the double-slit experiment to finitely many slits and to finitely many subsequent walls. Then the possibility space consists of all possible paths from the source *via* the walls to the detectors. Passing over to infinitely many walls with infinitely many slits leads to Feynman's path integral. For several other aspects of slit experiments, see Jansson [9], Chapter 4.

5. Hilbert's sixth problem

Hilbert's sixth problem [20] asks how to axiomatize those branches of physics in which mathematics, in the first rank the theory of probabilities, is prevalent. The aim is to treat physics by means of axioms, as in geometry.

DOI: http://dx.doi.org/10.5772/intechopen.1005070

Several different systems of axioms exist for probability theory. One of the most commonly used and well-known sets of axioms is the Kolmogorov axioms. These axioms are contained in our four principles *via* Born's rule for the outcomes. Our axiomatic approach to probability theory is similar to the axiomatic approach in geometry, where the foundational principles, such as Euclid's axioms, provide the basis for the development of geometric concepts and theorems. The axiomatic system in geometry consists of the following components:

- The primitives: points, lines, and planes.
- The axioms are statements about these primitives; for instance, two points are together incident with one line.
- The laws of logic.
- The theorems that are the logical consequences of the axioms.

According to Hilbert, primitive terms are empty shells or placeholders with no intrinsic properties. It means that instead of points, lines, and planes, we can also use the words windows, chairs, and houses. A concrete meaning of the primitives of a geometrical system yields a model of the axiomatic system, where all theorems are true statements in this model. Our possibility measure space may be viewed as an axiomatic probability theory in the sense of Hilbert's sixth problem, which is composed of the following components:

- The primitives: elementary possibilities, outcomes, and amplitudes.
- The axioms are statements about these primitives; for instance, each elementary possibility is contained in exactly one outcome.
- The laws of classical logic.
- The theorems, such as the inclusion-exclusion principle [9].

We shall consider several concrete models of our principles or axioms, thereunder Feynman's formulation of quantum mechanics in space-time, Wiener processes, and thermodynamics.

6. Consistency and symmetry

In the following, we prove the internal consistency of our probability theory, ensuring that it remains free from contradictions. Furthermore, we establish that our theory possesses a $U(1)$ symmetry, meaning that all probabilistic statements remain unchanged when the amplitudes associated with individual possibilities are transformed by a single element of the $U(1)$ group.

At first, we show that the probability amplitude φ_F is well-defined, that is, the amplitude should not depend on the partitioning of F. If F contains one element, there is nothing to prove. For two disjoint elements where $F = \bigcup\{F_1, F_2\}$, the amplitude

$\varphi_F = \varphi_{F_1} + \varphi_{F_2} = \varphi_{F_2} + \varphi_{F_1}$ is well-defined. In the case of three pairwise disjoint possibilities F_1, F_2, F_3, we partition $F = \bigcup\{F_1, F_2, F_3\}$ as follows:

$$F_1, F_2, F_3; \ \bigcup\{F_1, F_2\}, F_3; \ \bigcup\{F_1, F_3\}, F_2; \ \bigcup\{F_2, F_3\}, F_1. \tag{31}$$

Complex addition is associative and commutative. Hence, in each case, the first principle yields

$$\varphi_F = \varphi_{F_1} + \varphi_{F_2} + \varphi_{F_3}, \tag{32}$$

and φ_F is well-defined. Analogously, the same holds true when the partitioning consists of more than three elements:

$$\varphi_F = \sum_m \varphi_{F_m}. \tag{33}$$

The second principle, Born's rule, requires that the sum of the square of the magnitudes of all probability amplitudes that correspond to the outcomes is one. This is a simple normalization condition that can always be achieved.

Finally, due to Born's rule, multiplying all probability amplitudes with the same element $e^{i\phi} \in U(1)$ does not change the probabilities. Thus, our possibility measure space has a symmetry with respect to the fundamental symmetry group $U(1)$. It is well-known that quantum electrodynamics has a $U(1)$ gauge symmetry, justified by the fact that the absolute phase of the wave functions of electrons, photons, or other particles cannot be measured.

7. Reconstruction of quantum mechanics

In this section, we present a reconstruction of Feynman's quantum mechanics, rooted in the concept of path integrals. It is well-established that his theory is mathematical equivalent to both, Schrödinger's and Heisenberg's quantum formulations.

We introduce *Feynman's path integral* with the help of zigzag paths $x(t)$: Let a particle move from position x_a at time t_a to x_b at time t_b in space-time. The time is divided up into n smaller segments $t_a = t_0 < t_1 < \cdots < t_{n-1} < t_n = t_b$. All have the length $\varepsilon = (t_b - t_a)/n$.

The *possibility space* **P** contains finitely many zigzag paths from $a = (x_a, t_a)$ to $b = (x_b, t_b)$ where b varies in some subset B of the space-time. This subset may consist of various points where detectors are positioned. The *possibility algebra* **F** is the power set of **P**.

For fixed $b \in B$, the nonelementary possibility

$$F(b, a) = \{x(t) \in \mathbf{P} : x(t_a) = x_a, x(t_b) = x_b\} \in \mathbf{F} \tag{34}$$

defines an outcome. The sample space **O** consists of all sets $F(b, a)$ where b varies in B. They are pairwise disjoint and form a partitioning of **P**.

Let $c = (x_c, t_c) \in C$ be a space-time point such that $t_a < t_c < t_b$. We define the nonelementary possibility

$$F(b, c, a) = \{x(t) : x(t_a) = x_a, x(t_c) = x_c, x(t_b) = x_b\} \in \mathbf{F}. \tag{35}$$

Then

$$F(b,c,a) = F(b,c) \cap F(c,a), \tag{36}$$

where the sets on the right-hand side are defined as above. It follows that

$$F(b,a) = \bigcup_{c \in C} F(b,c,a). \tag{37}$$

Since the paths $x(t) \in \mathbf{P}$ are pairwise disjoint, the first principle implies Feynman's path integral:

$$\varphi(F(b,a)) = \sum_{x(t) \in F(b,a)} \varphi(\{x(t)\}). \tag{38}$$

The amplitude $\varphi(F(b,a))$ is well-known and also called *Green's kernel of motion*. Frequently, it is denoted by $K(b,a)$. Using Born's rule, we get the probability $Pr(b,a) = |K(b,a)|^2$ to move from a to b.

The *action* of a path is defined as the integral over its Lagrangian L

$$S(\{x(t)\}) = \int_{t_a}^{t_b} L(\dot{x}, x, t)dt. \tag{39}$$

We obtain the amplitude for the elementary possibilities with the third principle:

$$\varphi(\{x(t)\}) = \text{const} \quad \exp\left(\frac{i}{\hbar} S(\{x(t)\})\right), \tag{40}$$

We consider only zigzag paths. Thus, the action takes the form

$$S(\{x(t)\}) = \sum_{j=1}^{n} L\left(\frac{x_j - x_{j-1}}{\varepsilon}, \frac{x_j + x_{j-1}}{2}, \frac{t_j + t_{j-1}}{2}\right). \tag{41}$$

Formula (38) is the essence of the quantum formulation of Feynman. Now, we ask how to calculate the sum over all paths. We remember the Riemann integral of some function f, which is approximated in the form

$$\int_{x_a}^{x_b} f(x)dx \propto \sum_{j=0}^{n} f(x_j), \tag{42}$$

where the points x_j are equally spaced. This sum depends on the number n. In this form, a limit would not exist. But the normalization factor $\delta = (x_b - x_a)/n$ yields

$$\int_{x_a}^{x_b} f(x)dx = \lim_{\delta \to 0} \left(\delta \sum_{j=0}^{n} f(x_j)\right). \tag{43}$$

Similarly, we must introduce a normalization factor for the path integral. This is not trivial in concrete experiments.

Putting all together and taking the limit $\varepsilon = (t_b - t_a)/n \to 0$, *Feynman's path integral* Eq. (38) can be written as

$$K(b,a) = \lim_{\varepsilon \to 0} \frac{1}{A} \int \cdots \int \exp\left(\frac{i}{\hbar} S(\{x(t)\})\right) \frac{dx_1}{A} \cdots \frac{dx_{n-1}}{A}, \tag{44}$$

where A is a normalization constant depending on the Lagrangian.

The classical action is additive. Hence, Eq. (37), the first and fourth principle imply

$$S(b,a) = S(b,c) + S(c,a), \tag{45}$$

and it follows that

$$K(b,a) = \int_{x_c} K(b,c) K(c,a) dx_c. \tag{46}$$

More general, for $(n+1)$ points we get

$$K(b,a) = \int_{x_1} \int_{x_2} \cdots \int_{x_{n-1}} K(b, n-1) K(n-1, n-2) \cdots K(1,a) \; dx_1 dx_2 \cdots dx_{n-1} \tag{47}$$

where

$$K(j, j-1) = \frac{1}{A} \exp\left(\frac{i}{\hbar} \varepsilon L\left(\frac{x_j - x_{j-1}}{\varepsilon}, \frac{x_j + x_{j-1}}{2}, \frac{t_j + t_{j-1}}{2}\right)\right). \tag{48}$$

Now, we change the notation $x_b = x, t_b = t, x_a = y, t_a = s$. Then formula (46) can be written as a *wave function*, well-known in quantum theory:

$$\varphi(x,t) = \int K(x,t;y,s) \varphi(y,s) dy. \tag{49}$$

This formula says that the probability amplitude for the outcome of arriving at the point (x,t) is equal to the sum over all amplitudes to reach at (y,s) multiplied by the amplitude to move from (y,s) to (x,t).

In the prevailing formulation of quantum mechanics, the Schrödinger equation serves as the fundamental postulate. This equation can be derived from Eq. (49). We make an initial order approximation of the wave function with respect to the time interval ε, resulting in the formula:

$$\varphi(x, t+\varepsilon) = \frac{1}{A} \int \exp\left(\varepsilon \frac{i}{\hbar} L\left(\frac{x-y}{\varepsilon}, \frac{x+y}{2}, t\right)\right) \varphi(y,t) dy. \tag{50}$$

For example, in the special case of the Lagrangian $L = m\dot{x}^2 + V(x)$, we substitute $y = x + \mu$, integrate, and expand the resulting equation to first order in ε and second order in μ, yielding the *Schrödinger equation*

$$i\hbar \frac{\partial \varphi}{\partial t} = \frac{\hbar^2}{2m} \frac{\partial^2}{\partial x^2} \varphi + V\varphi. \tag{51}$$

DOI: http://dx.doi.org/10.5772/intechopen.1005070

The corresponding normalization constant turns out to be (see [21], Section 6)

$$A = \sqrt{\frac{2\pi\hbar\varepsilon i}{m}}. \tag{52}$$

Using our probability theory, we successfully reconstructed Feynman's formulation based on path integrals, ultimately deriving the Schrödinger equation. The process of quantization naturally emerges from this equation, establishing it as a direct outcome of our probabilistic framework. Furthermore, it is worth noting that quantization can also be derived directly from the path integral, as demonstrated by Kleinert (cf. [21], Sections 2.6 and 9.2). In contrast, classical probability theory does not imply the concept of quantization.

It can be shown that, in the limit case, the paths may exhibit continuity but lack differentiability throughout space-time. In other words, the velocity is discontinuous at every point.

The phase space path integral offers a broader perspective compared to the space-time path integral discussed above. It introduces momentum as a crucial parameter, establishing a connection between quantum mechanics and the Hamiltonian formalism.

We will not provide a detailed derivation of this path integral formulation. For an in-depth exploration of path integrals, including comprehensive information and references, readers are encouraged to consult Kleinert's monograph [21] and the related literature. Additional insights on this topic can be found in the works of Feynman (cf. [14, 16, 22]).

8. Diffusion and Wiener Integral

Readers with knowledge of statistical mechanics will readily observe a striking resemblance between Feynman's formulation and the concept of Brownian motion, where discretization mirrors the behavior of discrete-time random walks. In fact, the path integral formulation is closely related to the mathematical framework of Brownian motion. In this section, we aim to briefly outline the connections between quantum path integrals, Brownian motion, diffusion processes, and the Wiener integral. For a more comprehensive exploration, readers are encouraged to consult Zeidler's book [23], Chapter 11 and explore the relevant literature.

The heat equation is defined as a partial differential equation, specifically addressing an initial value problem with an initial time parameter s:

$$\frac{\partial\varphi(x,t)}{\partial t} - -\kappa\frac{\partial^2}{\partial x^2}\varphi(x,t) \quad V(x)\varphi(x,t), \quad t \geq s, \quad \psi(x,s) = \psi_0(x). \tag{53}$$

In addition to its wide-ranging applications in scientific domains such as probability theory, financial mathematics, and image analysis, this equation provides a fundamental description of heat propagation within an isotropic and homogeneous medium. In this context, the variable $\varphi(x,t)$ represents the temperature at a specific spatial point, denoted by x, and at a particular moment in time, by t. Furthermore, this equation serves as a diffusion equation when applied to a mass density, with $\varphi(x,t)$ representing this density. At the microscopic level, diffusion is intimately connected to Brownian motion, which characterizes the stochastic and random movement of microscopic particles within gases or liquids.

In the book of Zeidler [23], Section 11.8, it is proved that its solution is

$$\varphi(x,t) = \int K(x,t;y,s)\varphi_0(y)dy, \tag{54}$$

where the heat kernel has the form

$$K(x,t;y,s) = \lim_{\varepsilon\to 0} \frac{1}{A}\int \cdots \int \exp(-S(\{x(t)\})) \frac{dx_1}{A} \cdots \frac{dx_{n-1}}{A}. \tag{55}$$

The symbol S represents the discrete action associated with a linear zigzag path, denoted as $x(t) = (x(t_i))$. For a Lagrangian, which is defined as the difference between the kinetic energy and the potential energy V, we have the expression:

$$S(\{x(t)\}) = \sum_{j=1}^{n} \frac{1}{4\kappa}\left(\frac{x_j - x_{j-1}}{\varepsilon}\right)^2 + V(x_j)\varepsilon. \tag{56}$$

We take the same discretization as in Section 7. The resulting normalization constant for points in the three-dimensional position space is

$$A = (4\pi\kappa\varepsilon)^{3/2}. \tag{57}$$

Much like Feynman's path integral, the heat kernel denoted as $K(x,t;y,s)$ represents the summation over all possible paths connecting the initial point y to the final point x.

The path integral in Eq. (55), which is also known as a *Wiener integral*, possesses a well-defined and rigorous interpretation as a classical measure within the realm of continuous functions ([24], Vol. II, Section X.11).

9. Thermodynamics

It is an important touchstone to reconstruct thermodynamics using our probability framework. For a more in-depth exploration of this reconstruction, we refer to Jansson ([9], Chapter 5). For those seeking a comprehensive introduction to the theory of thermodynamics, we recommend Penrose ([25], Chapter 27), and Ben-Naim [26, 27].

In thermodynamics, we often deal with an enormous number of constituents. To illustrate, just one mole of molecules corresponds to Avogadro's number, which is approximately on the order of 10^{23}. Consequently, thermodynamics fundamentally operates as a statistical theory.

The large collections of constituents are described in terms of *microstates*, where each constituent possesses attributes such as position, momentum, or energy. A microstate represents a specific configuration of a system where all microscopic variables are precisely determined. Microstates are distinct possibilities; they either occur or do not in the present, but two or more microstates cannot coexist simultaneously.

Macrostates, on the other hand, pertain to the overall thermodynamic system. These macrostates are characterized by a small set of macroscopic variables, such as the total energy E, pressure P, volume V, temperature T, or the total number N of molecules. Throughout the following discussion, we will use M to denote a macrostate and μ to denote a microstate.

The number of microstates, each representing precise configurations with exact microscopic values, can be immensely large. In contrast, a macrostate is defined by the fixation of a small number of macroscopic variables. Each macrostate encompasses a multitude of microstates, often referred to as *accessible* microstates. The *multiplicity* of a given *macrostate* denoted as M is the number of its accessible microstates and is represented as $\Omega(M)$. The *total multiplicity*, denoted as Ω_{tot}, is the sum of all the multiplicities $\Omega(M)$.

Macrostates are measurable in contrast to microstates, and they effectively partition the set of all microstates within the system.

The foundational principle of statistical thermodynamics asserts that *all microstates within a system are equiprobable*. As a consequence of this principle, the probability associated with a macrostate M is determined by the ratio of the multiplicity of that macrostate to the total multiplicity:

$$Pr(M) = \frac{\Omega(M)}{\Omega_{tot}}. \tag{58}$$

The obvious way to merge statistical thermodynamics with our probability framework is to identify the microstates μ as the elementary possibilities $p \in \mathbf{P}$ and to associate the macrostates M, as measurable states, to the outcome $F \in \mathbf{O}$.

Now, we can reevaluate the probabilities associated with macrostates Eq. (58) using our probabilistic framework. Our *third principle* posits that all elementary possibilities contribute equally in magnitude, meaning that the microstates μ can be expressed with amplitudes as follows:

$$\varphi_\mu = \text{const } e^{\frac{i}{\hbar}S(\mu)}. \tag{59}$$

Without further knowledge about the actions of the constituents, it is reasonable to assume that the action $S(\mu)$ is uniformly zero for all microstates. This choice renders the exponential term equal to one, indicating no interaction or interference. Furthermore, we set:

$$\text{const} = \frac{1}{\sqrt{\Omega_{tot}}\sqrt{\Omega(M)}}. \tag{60}$$

Then

$$\varphi_\mu = \frac{1}{\sqrt{\Omega_{tot}}\sqrt{\Omega(M)}} \cdot 1. \tag{61}$$

Since the microstates are mutually exclusive, we can invoke the first principle. Consequently, the probability amplitude of a macrostate M takes the form:

$$\varphi_M = \sum_{\mu \in M} \varphi_\mu = \Omega(M) \frac{1}{\sqrt{\Omega_{tot}}\sqrt{\Omega(M)}} = \sqrt{\frac{\Omega(M)}{\Omega_{tot}}}. \tag{62}$$

Following Born's rule, we obtain the classical probabilities for the outcomes as expressed in Eq. (58) when we compute the squared magnitude of probability amplitudes.

It is important to note that in our derivation, we did not employ the thermodynamic principle of indifference. Rather, our approach hinges on setting the action of all elementary possibilities (microstates) to zero. This choice pertains to the experimental setup rather than making a statement about probabilities.

Einstein writes about thermodynamics:

> *A theory is the more impressive the greater the simplicity of its premises is, the more different kinds of things it relates, and the more extended is its area of applicability. Therefore the deep impression which classical thermodynamics made upon me. It is the only physical theory of universal content concerning which I am convinced that within the framework of the applicability of its basic concepts, it will never be overthrown.*
> Albert Einstein, autobiographical notes (1946)

with our probabilistic framework, we have covered many applications and reconstructed theories in addition to statistical thermodynamics.

We have introduced a probability theory describing future events. The future is timeless. Not surprisingly, the foundational theory of statistical thermodynamics, almost universally applicable, is also inherently timeless, as highlighted by the work of Ben-Naim [26]. The second law of thermodynamics and the concept of entropy are independent of time. This perspective aligns with the notion of "physics without time" advocated by some physicists, see Rovelli [28].

10. Conclusion

John Wheeler, as mentioned by Ballentine [29], contended that true comprehension of quantum theory demands the ability to encapsulate it within a single, readily understandable statement. Our succinct statement is as follows:

> *Quantum theory can be reconstructed through a simple probability framework that characterizes future events in terms of possibilities and outcomes, employing classical logic, straightforward set theory, and complex numbers.*

Our approach to quantum theory departs from conventional quantum mechanics in several key aspects. Moreover, our framework allows for the reconstruction of classical statistical mechanics and thermodynamics.

The theory appears to be simple enough to be taught even in schools, similar to Kolmogorov's theory of probability.

Theories and interpretations can significantly influence techniques and engineering practices. Quantum information theory provides insights into communication systems, data compression, and cryptography, essential in modern engineering practices such as telecommunications, information technology, and cybersecurity. The two fundamental properties of quantum mechanics: superposition (see Section 3) and entanglement (see Jansson [7], Section 4, where the theory of special relativity is reconstructed in a six-dimensional Euclidean space), receive not only a new interpretation but also a new mathematical framework. I hope this will lead to new insights in quantum information science.

Acknowledgments

I am grateful to Otfried Ischebeck, Frerich Keil, and Thomas Künemund for their critical reading of the manuscript and their suggestions. An enlarged version of this chapter was published as a preprint and stated under the title "Conceptual basis of probability and quantum information theory" on the preprint server "https://tore.tuhh.de" in the year 2022.

Additional information

https://www.tuhh.de/ti3/jansson/

Author details

Christian Jansson
Institute for Reliable Computing, Hamburg University of Technology, Hamburg, Germany

*Address all correspondence to: jansson@tu-harburg.de

References

[1] Shafer G, Vovk V. The sources of Kolmogorov's Grundbegriffe. Statistical Science. 2006;**21**(1):70-98

[2] von Weizsäcker. The Structure of Physics (Original 1985). Netherlands: Springer; 2006

[3] Rudas T. Handbook of Probability. SAGE Publications; 2008

[4] Weinberg S. The trouble with quantum mechanics. New York Review of Books. 2017;**64**(1):51-53

[5] Fuchs C. Quantum Mechanics as Quantum Information (and Only a Little More). 2002.arXiv preprintquant-ph/ 02050392002

[6] Omnes R. The Interpretation of Quantum Mechanics. Vol. 102. Princeton University Press; 1994

[7] Jansson C. Quantum Information Theory for Engineers: An Interpretative Approach. Hamburg University of Technology; 2017. DOI: 10.15480/882.1441

[8] Jansson C. Quantum Information Theory for Engineers: Free Climbing through Physics and Probability. Hamburg University of Technology; 2019. DOI: 10.15480/882.2285

[9] Jansson C. A Unified Treatment of Classical Probability, Thermodynamics, and Quantum Information Theory. Hamburg University of Technology; 2021. DOI: 10.15480/ 882.3770

[10] Jansson C. Conceptual Basis of Probability and Quantum Information Theory. Hamburg University of Technology; 2022. DOI: 10.15480/ 882.4590

[11] Zee A. Fearful Symmetry: The Search for Beauty in Modern Physics. Vol. 48. Princeton University Press; 2015

[12] Susskind L, Friedmann A. Quantum Mechanics: The Theoretical Minimum. Basic Books; 2014

[13] Fuchs CA, Peres A. Quantum theory needs no 'interpretation'. Physics Today. 2000;**53**(3):70-71

[14] Feynman RP, Leighton RB, Sands M. The Feynman Lectures on Physics. Addison Wesley; Later Printing edition; 1971; 1963

[15] Wood C. Imaginary Numbers May be Essential for Describing Reality. Quanta Magazine; 2002

[16] Feynman RP, Hibbs AR. Path Integrals and Quantum Mechanics. New York: McGraw; 1995

[17] Smolin L. Einstein's Unfinished Revolution: The Search for What Lies beyond the Quantum. Penguin; 2019

[18] Cease RP. The most beautiful experiment. Physics World. 2002;**15**(9): 19-20

[19] Penrose R. Fashion, Faith and Fantasy. Princeton and Oxford: Princeton University Press; 2016

[20] Hilbert D. Mathematical problems. In: Mathematics. Chapman and Hall/ CRC; 2019. pp. 273-278

[21] Kleinert H. Path Integrals in Quantum Mechanics, Statistics, Polymer Physics, and Financial Markets. World Scientific; 2009

[22] Feynman RP. Space-time approach to non-relativistic quantum mechanics. Reviews of Modern Physics. 1948;**20**:2

[23] Zeidler E. Quantum Field Theory I: Basics in Mathematics and Physics. Berlin, Heidelberg: Springer-Verlag; 2006

[24] Reed M, Simon B. Methods of Modern Mathematical Physics. New York: Academic Press; 1972

[25] Penrose R. The Road to Reality: A Complete Guide to the Laws of the Universe. New York: Bodley Head; 2005

[26] Ben-Naim A. The Briefest History of Time. World Scientific; 2016

[27] Ben-Naim A. Time's Arrow (?) the Timeless Nature of Entropy and the Second Law of Thermodynamics. New Jersey: Princeton University Press; 2018

[28] Rovelli C. The Order of Time. Singapore, New Jersey, London, Hong Kong: Pinguin Books; 2018

[29] Ballentine LE. Quantum Mechanics: A Modern Development. World Scientific; 2014